# THE
# LEADING EDGE

WALTER J. BOYNE

ARTABRAS · PUBLISHERS · NEW YORK

Published by arrangement with Stewart, Tabori & Chang, Inc.
Artabras edition 1991 (updated reprint of original edition)

Library of Congress Cataloging-in-Publication Data
Boyne, Walter J., 1929–
    The leading edge / Walter J. Boyne.—1st Artabras ed., rev. and updated.
        p.    cm.
    Originally published: New York: Stewart, Tabori & Chang, 1986.
    Includes index.
    ISBN 0-89660-020-3
    1. Airplanes—Technological innovations—History.   I. Title.
TL671.2.B633   1991
629. 13—dc20                                      90-47475
                                                       CIP

**PAGE 1:**
"The missile with a man in it," Kelly Johnson's Lockheed F-104.

**PAGE 2:**
The twin engine Beech D-18, for many years the most advanced executive transport in the world.

**BELOW RIGHT:**
Charles "Daredevil" Hamilton.

**DESIGN**
J. C. Suarès
Gordon Harris

## ACKNOWLEDGMENTS

Many persons and individuals have contributed enormously to this book, with advice, photos and review. I want to especially thank Janis Davidson, whose own inveterate fascination with aviation caused her to work in her free time to make the book better; Jay Miller, who took time off from his enormous duties to help so many times; Richard S. Allen, whose knowledge of old aircraft is incredible; R. Capel and Jean Salis; Kirby Harrison; Gene Chase and his EAA group; Art Davis; Stefan Nicolaou; Robert S. DeGroat; Fred Hayward of the magnificent Shuttleworth Collection; Mike Long; Ed Phillips; Chris Wamsley of Rockwell International; Dick Milligan of Grumman; David Lee, Imperial War Museum; Gerald Balzer; Commander Dick Schram, U.S. Navy; Dr. John Tanner, RAF Museum; Ralph Wheeler, General Electric; George Weiss, Boeing; Don MacDonald, of McDonnell Douglas; the late Willard Custer; Leo Schefer, British Aerospace; Ted LeFevre, General Dynamics; Scott Crossfield; Budd Davisson; Jacques de Spoelberch, and many, many others. One last comment: the community of aviation writers is an amazing group; always cooperative, invariably generous, they always respond to requests with alacrity. It is an honor to be a part of their fraternity.

Walter J. Boyne
Alexandria, Virginia
May 28, 1986

# CONTENTS

# INTRODUCTION

For more than eighty years, aviation has treated us to a never-ending stream of new ideas for protecting our shores, shrinking distances, and making more homogeneous our world. Each new idea has been the leading edge of a particular technology —a shining sharp blade cleaving the way to the future, turning older ideas aside. It began with the Wright brothers and goes on at this moment in a host of forms, ranging from the next generation of hypersonic airliners to gossamer-winged human-powered vehicles.

Appreciation for the combination of leadership and change is a peculiarly human attribute, calling as it does for continual improvement and endless competition. In an ideal world, every breakthrough would lead directly to new successes, new profits, and perhaps even happiness for its finder. In fact, however, the leading edge at any moment may not be perceived, may succumb to even newer ideas, or may fail for some time to provide a measurable (and cost-effective) improvement over the old. Still, any genuine leading edge will eventually take its place at the forefront of technology.

## WHAT IS A LEADING EDGE?

The term *leading edge* may now be more familiar in science, particularly with reference to computers and electronics, than it ever was in aviation. Here,

**OPPOSITE:**
This Curtiss biplane from the Owls Head Transportation Museum has a tricycle landing gear which made ground handling easy but added weight, drag, and lengthened the landing roll.

**OVERLEAF:**
By 1911, the basic Wright design had come a long way. Here the Wright EX biplane takes off from Sheepshead Bay, Long Island.

it refers to any device or system used in aviation that extends the capability of an aircraft to its farthest limits. The limits can be of any sort: speed, altitude, range, maneuverability, safety, or some combination of these. But there are human complications to this formula, wonderful diversions from logic or leaps to truth that depend upon the individuals involved. The inventor's personality, for example, is intimately involved in the early success—or perhaps premature failure—of a leading edge, as is the inventor's ability to place the invention in the right context. Another human complication is less evident now than in the first four decades of flight, before test pilots were also trained scientists. In the early years, the pilot factor was as important as the invention, and a negative report from a test pilot could stymie progress for years.

Of course, some leading edges led nowhere, being overtaken by other inventions or events. The inventor or designer enjoyed a golden moment of anticipation about being at the fore, only to find that time or science had passed him by.

In looking back, we see that some great ideas occurred too soon—the tricycle gear was thirty-five years early, the jet aircraft thirty—and the aviation world simply waited until the time was right. In other instances, the great idea occurred when the world was waiting for it, as with the structural use of modern composite materials.

## The Movers and Shakers

Characteristically, the men and women who create leading edges are pushed by necessity and pulled by circumstance. In the early 1930s, when the science of instrument flight was being pushed by the burgeoning need for mass air transportation, it became necessary to create, among other devices, de-icing equipment. Simultaneously, and independently, aircraft were changing from the traditional biplane configuration (on which de-icing equipment could not be fitted) to the now-familiar cantilever monoplane arrangement, which lent itself in a variety of ways to de-icing gear. The two innovations were married in the Douglas DC-3, and the previously insurmountable problem of removing or preventing ice accretion on airfoils became solvable through procedures and technique. This happy confluence of improvements occurs time and again.

An engaging aspect of taking a retrospective view of the leading edge in aviation consists in witnessing how often people of genius perceived the possibility of a great new advance before a genuine need for it existed. Thus, in 1917, J. V. Martin placed the first practical retractable landing gear on an aircraft that would not fly, and in 1924 Walter Barling created a gigantic six-engine three-wing bomber that could carry a large load of bombs or could fly above 3,000 feet, but could not do both. Professor Henri Coanda, whose scientific work was impeccable, designed and built a jet aircraft in 1910; it, like Martin's Kitten, was superbly built and technically advanced—and could not fly.

Sometimes, the leading edge of an idea distorts the thinking of its inventor to the exclusion of common sense. More often, however, the inventor is the only person with a comprehensive view of the entire goal and is the only one with the assurance to carry it through. Dr. Paul MacCready became interested in the problem of human-powered flight, a subject that was under intensive investigation in many countries of the world. The traditional approach was to build the most elegant possible version of an ultralightweight glider, employing complex structures to achieve the necessary strengths at

**RIGHT:**
The instrument panel of Charles Lindbergh's *Spirit of St. Louis*—spartan, austere, but just sufficient.
**OPPOSITE:**
As aircraft grew in sophistication, so did their control systems.

**OVERLEAF:**
Sometimes a magnificent leading edge is too expensive to operate—such was the case with these Mach 2 Convair B-58 bombers, here awaiting dismantling.

minimum weight. MacCready took a different tack, using highly unconventional structures in an aesthetically dreadful airframe. But MacCready had the vision, and his crude-appearing aircraft was light, strong, and easy to repair. The MacCready team could crash one day and be flying the next, while a crash in the competitors' vehicles meant weeks or months of repair work. MacCready saw not only what the goal was, but what could and must be done to attain it.

The avant-garde artist breaking new ground in a medium and the inventor pushing forward an idea that will become the leading edge have much in common. They both must commit themselves totally to their *idée fixe* and become oblivious to the naysayers. They must devote themselves entirely to their work, and they must risk their lives, property, and reputations on their efforts.

*The Leading Edge* will take a look over time at the products of these artist-inventors of aviation, these men and women of vision whose determination to push aviation forward has made all the risks worthwhile. We will find that the mechanical devices responsible for advancing the science of

aviation have a beauty of their own, an inherent logic that qualifies them not only as inventions of genius but as works of art, and we will see that behind each one lies a veritable soap opera of human drama.

## TURBULENT IDEAS TODAY

As aviation reaches for its ninth decade, have all the good ideas been explored? Have all its leading edges grown just a little rusty? Not at all. Aviation today is bursting with new ideas, is being driven by demands for new achievements, and is bristling with capabilities never possible before.

At least partial responsibility for this situation lies in the fact that aviation is also in perilous trouble, faced with difficulties of a magnitude and complexity never before imagined. General aviation, that wellspring of American flying hopes, is moribund, condemned to death by an indifferent population, rising costs, and the specter of outrageous awards in liability suits. Recently, a seven-year-old light plane was involved in a fatal accident in which the problem was found to be a faulty and worn seat-adjustment fastener. The plaintiff won $3 million in compensatory damages and $9 million in punitive damages—for an airplane that had left the factory seven years before. The price tag of the de minimis light plane, the flight school trainer, is about $45,000, of which $19,000 is a provision for legal liability.

Even participation in military aviation is dwindling as a result of spiraling costs. In 1944, a B-29 bomber cost $509,465 and a P-51 fighter cost $50,985; the inventory strength of aircraft was 78,757. Now an F-15 fighter costs $28.5 million, a B-1B bomber costs almost ten times that, and the Air Force's inventory strength has been reduced to 7,300 planes. The experts are only partly joking when they predict that the air force of the future will consist of a single $5 trillion airplane.

**OPPOSITE:**
Today the future looks beyond supersonic to hypersonic vehicles that will change the world as surely as the Wrights' fragile-appearing biplane did.

Commercial air transportation is hardly immune. The effects of deregulation have been staggering: long-established companies saddled with substantial union contracts found themselves in trouble when faced in a deregulated marketplace with the host of new airlines that sprang up like mushrooms after a rain. When the new airlines cut costs with lower salaries and reduced amenities, some of the older carriers were unable to adjust to the keen competition and simply disappeared.

How can it be, then, that there are more leading edges than ever before—new configurations, new structures, new devices, and the capability to create even more of these through the powerful mathematics of computers? How can this time of rising cost, increased regulations, and diminished interest be a time when the most fascinating airplanes ever built are being demonstrated in a vast variety of shapes, sizes, and functions?

Part of it may be explained by the fact that aviation has always moved by great leaps and bounds, surging ahead when a confluence of ideas makes progress possible, and then stopping for a while to digest the advances and prepare for the next round.

Yet the real explanation is that science, sociology, and need have all come together at a single point. On the one hand, the enormous powers of computers are enabling engine and airframe manufacturers to plan new military and commercial aircraft of incredible sophistication and refinement; on the other, new ideas are pouring forth from the individual pilots who want to fly badly enough to build their own airplanes. An added spice in this frothing brew of innovation is the ultralight movement, creating a whole band of new flyers at minimum expense. Technique and materials are being intermixed through these three elements, and the result constitutes the leading edge of the aircraft of tomorrow.

# 1. THE EARLIEST EDGE

Imagine the scientific furor that would ensue if a team of archaeologists working in Africa were to announce, not that they had found yet another shard of jawbone belonging to yet another ancestor of homo sapiens, but instead that they had located the very place where man first discovered fire! Suppose they were to offer incontrovertible evidence that the blackened ashes and charred wood they had found marked absolutely the first instance in history where human beings had deliberately created a fire. What an uproar there would be in academe! Halls would be hired for years in advance to handle the proliferation of symposia. Public broadcasting stations would be deluged with shows covering every detail of the discovery, complete with interviews of everyone remotely concerned. Exactly the same whirlwind of attention would surround the verification of the first wheel. Teams of scientists would invade Syria or Afghanistan or China or wherever this first-round attempt to improve man's lot was found.

Among all the monumental firsts of human life on earth—achievements comparable to the first man-made fire or the first wheel—only one is readily at hand as an artifact, visible to all. Suspended in the center of the National Air and Space Museum, within slam-dunk range of an average basketball player, it is the Wright airplane (now lovingly called the Kitty Hawk Flyer), which made the first

**OPPOSITE:**
Popularly called ''the Flyer,'' the first successful man-carrying aircraft was flown at Kitty Hawk, North Carolina, on December 17, 1903.

controlled flight on December 17, 1903. But instead of inspiring awe, as would the remains of the first fire or the first wheel, the Flyer more commonly elicits affectionate amusement. It is too far from us to be remembered, too close to us to be revered.

We need only leaf through the magazines of the first fifteen years of the twentieth century, however, to see the grip it had on people's minds. Although almost no one had seen an airplane fly and few newspapers had published actual photographs of such an occurrence, the Wrights' biplane, with its forward elevator and rear vertical rudders, became a familiar sight. It appeared on kites, on calendars, on cookbooks, on the covers of popular songs, and on the frontispieces of children's books of fiction. The airplane's shape, far more than its accomplishments or potential, was firmly established in the American psyche. Oddly enough, that was perhaps the only place the airplane was established at all, for it was eschewed by the military, ignored by science, and almost repudiated by business.

Despite all we hear of the influence of early aviation, the fact is that both government and business in America rejected the first airplane in every practical sense, relegating it to carnival performances on county fair circuits. Practical proof of this appears in the business journals of the day, such as *Iron Age* and *Printer's Ink*. There the pounding pulse of American business is recorded in stories

on the importance of the automobile, of insurance, of agriculture—but never of airplanes. The airplane became a kind of instant legend, striking evidence of what Americans could do; but there was no evidence of what airplanes could do for America.

Part of the problem was the gossamer appearance of the Flyer. Even today, the single adjective most often used to describe the 1903 Wright Flyer is "fragile," but this is as erroneous now as it was when the airplane flew. Used as we are to the tough metal hulls and stout landing gear of jet liners, the wood and fabric of the Flyer does not seem particularly robust. And yet it was and is immensely sturdy, a brilliant arrangement of thin wooden spars and tightly strung wires creating a structure of enormous resilience and strength.

The Flyer is like this because the Wright brothers intended it to be so, just as they realized their intention for every other element of its design. The Wrights not only built the first airplane, they did it in a scientific manner that was not matched by other experimenters for a decade or more. If all the elements of the Wright Flyer are properly understood, the preferred adjective is not "fragile," but "harmonious," "appropriate," or "balanced"—words that describe the almost perfect way in which each element of the Flyer did its job.

## TWO MEN OF GENIUS AGAINST THE WORLD

By studying the Wright Flyer, we can begin to appreciate the meticulous engineering of the two young men from Dayton on their obsession, the flying machine. It has *just* enough wing area, *just* enough power, *just* enough control; this exact sufficiency was obtained only by stretching the limits of the Wright brothers' well-matched minds—and flying skills—to achieve far more than any existing engineering manual said was possible.

A number of characteristics of what would today be called the "Wright Brothers' Test Program" reinforce the idea that their Kitty Hawk Flyer was a milestone of aeronautical engineering achievement. In the first place, the Wrights labored in a field whose most accomplished practitioner, Otto

Wilbur Wright (*top*), and Orville Wright.

Lilienthal, had created a basic body of error in establishing the relationship of wing shape to lift. Despite his miscalculations and the fundamental unsoundness of his approach, Lilienthal was amazingly successful with his creations. Pictures of him hang gliding, his legs dangling beneath one of his flat-winged creations, inspired people around the world, including the Wrights.

## THE FATAL LILIENTHAL ERRORS

Lilienthal was a victim of at least three deficiencies in scientific insight. First, most important, and also most distinct from the Wrights' approach was his longstanding failure to perceive the need for three-axis control; he attempted instead to control his gliders by shifting the weight of his body. (Later he had some designs ready to test that would have incorporated a mechanical control system.) Second and more closely related to the formulaic errors noted above was his use of incorrect data to complete lift. After ignoring the possibility of mechanical control, he painted himself into a scientific corner by depending on airfoils and wing shapes that comported with his inaccurate formulas; these in turn forced Lilienthal to build wings that were so large that he could no longer control his craft by shifting his weight.

The third failure in his thinking did not affect the success of his practical achievements because he died before he could test his idea for power. However, he had studied birds and planned to add power to his gliders by using pinionlike wing extensions to generate propulsive power. It would never have worked.

On August 9, 1896, after some 2,500 gliding flights, Lilienthal was killed in a crash. He became the first hang glider victim since Icarus because his vehicles were designed according to the errors he had made in his formulas for flight.

## THE REWARDS OF DISBELIEF

The Wright brothers used Lilienthal's tables as a basis for their 1901 glider. When the glider failed to deliver the performance it should have, they were immensely discouraged, not daring to attribute its failure to mistakes by the master in the field. And yet, back in Ohio, they applied the crystal-clear reasoning of their "conjoined" brains to the project and found by experimentation that Lilienthal was lamentably wrong. In essence, Lilienthal had not correctly estimated the amount of lift generated by a wing, and yet he had established tables (upon which his and the experiments of others were based) using the incorrect information. The practical result was that he and other inventors could not obtain the lift they expected from given wing areas. The Wrights were iconoclasts enough to disbelieve the established tables, and they were scientists enough to establish their own tables based on results obtained with their own homemade wind tunnel—the first in North America.

Iconoclasts they may have been, yet their 1902 glider proved them to be entirely correct; at this point they began to learn to glide, making hundreds of flights in their season at Kitty Hawk. Their scientific deviationism was probably possible only because of the devil's advocate method of argument they used on each other to sort out the facts. Wilbur and Orville Wright were different in appearance and manner, but they worked together in a harmony rare even between identical twins. Wilbur was four years older but had been ill at a time when Orville was entering young manhood. The illness seems to have erased differences in age and outlook, and the Wrights functioned as one for the rest of their lives together.

The insight telling them that they must *learn* to fly was strangely denied to almost every other would-be practitioner. Harry Coombs, author of the brilliant *Kill Devil Hill*, points out that there were some 1.8 billion people on the earth in 1903, and of these only the two Wright brothers were pursuing a correct path of inquiry to solve the problem of flight. One endemic error was the belief that an aircraft in the air would be very much like a ship in the water, meaning that a helmsman could turn a rudder to the right or left or turn a horizontal surface up or down and thereby, in Charles Gibbs-

Otto Lilienthal.

**OVERLEAF:**
A genius whose gliding flights inspired would-be aviators around the world, Lilienthal built an artificial hill so that he could always leap directly into the wind.

Smith's words, "chauffeur the aircraft" about.

The Wrights, however, were vastly impressed by the versatility of bird flight—the darting, the wheeling, and the ever-changing control that birds exercised—and they believed this was the model to use. To achieve it, they designed an aircraft that did not have the basic stability of a drifting boat, but instead required a skilled hand on controls governing it about all three axes of flight. The controls were as simple and practical as the other elements of the Wright design. Lateral control was maintained by the flexible wings, which could be warped to produce a turn. Horizontal stability was provided by the elevator mounted in front, a disposition they correctly believed would alleviate the stall and crash sequence that had killed Otto Lilienthal.

The Wrights' success in establishing control in the air was due to a combination of intuition and practical experience. In their 1901 glider, they had made glides of almost 400 feet, but the craft had exhibited an alarming tendency to slue around when turned and to sideslip into the ground. At their low flying speeds, the additional drag induced by the positively warped wing tip caused it to slow down, and they had not yet learned how to offset this with a coordinated movement of the rudder. Unlike many would-be practitioners, they had not expected to make flat turns in the air, but to bank just as they banked around curves on their bicycles. The sideslipping phenomenon intruded as one more problem to be solved.

The solutions came in the long, cold Ohio winter. The Wrights returned to Kitty Hawk in August 1902; they flew more than a thousand glider flights by October of that year and solved the problem of control by integrating the action of the rudder with the warping of the wings. (This coupling solution was dropped by them later, when they learned to operate the rudder manually to achieve the same effect, but it appeared long afterward in many other aircraft—most notably the Ercoupe light plane.)

The success of the 1902 glider convinced the Wrights that they were ready for powered flight. The wing warping worked well in conjunction with

the vertical rudder surface for lateral control, both movements being operated by shifting the pilot's hip in a leather saddle device. The forward elevator was really too responsive, offering precise control and serving to prevent the sort of stall that had killed Lilienthal. All that remained was to install an engine and propellers, with a suitable transmission arrangement.

## SCIENTIFIC COMPETITION

The Wrights were now already a decade ahead of all competitors. They had funded the entire program themselves, using the proceeds from their

It was a curious and fortunate coincidence that aviation and motion-picture photography came into being almost simultaneously. Here a Wright biplane makes a "high-speed pass" over a row of cameramen.

successful bicycle business to buy the materials and to travel to Kitty Hawk. The fact that they were successful businessmen had only one drawback, a historical one. For years after their achievement, it was common practice to describe them as "bicycle mechanics," home-grown American craftsmen who had somehow put together the correct formula to fly. Nothing could be farther from the truth, and their scientific achievement is all the more notable when compared to the efforts of a great scientist, Samuel Pierpont Langley, the third Secretary of the Smithsonian Institution. Although neither party knew it, the Wrights and Langley were seemingly in a neck-and-neck race to be the first to fly. It is instructive to examine Langley's approach and see how close to (or far from) succeeding he was.

A distinguished scientist whose contributions to his own field of astrophysics were enormous, Langley approached the problem of flight in a strangely intuitive way, using an imprecise mixture of exactitude and guesswork. Despite his formal training, something in the nature of flight elicited a response in him that he would have immediately recognized as nonscientific in his own field. His first efforts at building model aircraft were nonetheless markedly successful. In 1892, he began experimenting with steam-powered tandem-wing aircraft (a configuration that has inspired many other engineers, even to this day), and on May 6, 1896, his unmanned model Aerodrome No. 5 made the first successful flight of any heavier-than-air craft.

Precisely fashioned and launched from an ingenious catapult atop a houseboat on the Potomac River near Quantico, Virginia, the Aerodrome flew in stately circles to a height of almost 100 feet, and then glided steadily back down to the water, having covered an estimated distance of 3,300 feet. Even though it had no alighting gear, it suffered no damage and was able to make an even more impressive flight later in the day.

Aerodrome (the unfortunate term was coined by Langley, applied to his own creations, and then mercifully dropped) No. 6 also flew successfully, and Langley retired temporarily from the flight arena, confident that he had demonstrated the configuration and technique necessary for manned flight.

The short, brutal war with Spain in 1898 excited the War Department's interest in heavier-than-air craft for the first time. It appropriated the unheard-of sum of $50,000 (perhaps ten times the total investment the Wrights had made prior to 1903) for Langley's use. Enthusiastic and confident, Langley embarked upon an ambitious program to create a full-size, man-carrying Aerodrome. The program was to be distinguished by exquisite craftsmanship and egregious scientific blunders. In an almost reverse image of the Wrights' thinking, Langley

**ABOVE:**
Samuel Pierpont Langley.
**RIGHT:**
In 1914, Glenn Curtiss modified the Aerodrome as a part of his patent fights with the Wrights. Langley, a scientist, would not have approved the ploy (*top*). Curtiss on the land-bound Langley, at the 1914 trials of the modified Aerodrome (*center*). The Aerodrome sitting on its houseboat, near Quantico, Virginia (*bottom*).

24

assiduously ignored the central problems—problems so enormous that his program was doomed to failure from the start.

In crafting his man-carrying Aerodrome, Langley simply scaled up his highly successful models by a factor of about four, ignoring the fact that materials, structures, and stresses do not scale up equivalently. He persisted in using a catapult launch over water idea, despite the increase in mass, stress, and speed inherent in the larger aircraft. He neglected to fit the Aerodrome with any undercarriage at all. There were two disadvantages to this approach that should have been apparent to Langley and must have been screamingly obvious to the pilot designee, Charles Manly. First, the pilot of the Aerodrome was denied any opportunity to gain familiarity with the aircraft through taxi tests. Second, every time the Aerodrome landed in the water, its pilot would be submerged beneath the vehicle; and if by chance it was forced to land on the ground, the pilot would have been scrubbed away by the impact. Manly must have been incredibly brave.

Even more troublesome from a scientific point of view was the fact that Langley, like everyone else but the Wrights, neglected the problem of three-axis control and therefore had to depend on the Aerodrome's stability to cause it to fly gently enough for the pilot to control with the almost pathetic control surfaces Langley provided.

In sum, then, Langley launched an aircraft that had not been flown and could not be controlled, using a catapult that had never been used to launch a full-size vehicle, a pilot who had never even glided, and an engine that had never flown. The permutational certainty of failure was reinforced by basic aerodynamic and structural flaws that ensured instant destruction upon launch.

The one bright spot in the entire endeavor, aside from the flawless workmanship involved, was the engine. Charles Manly, the brave man who twice tried to fly the Aerodrome and twice landed ignominiously in the Potomac, had converted a rotary engine designed by Stephen M. Balzer into a magnificent, water-cooled radial of 52 horsepower. It

Charles Matthews Manly.

is just a little exaggerated to say that if the Wrights had had the Balzer/Manly engine, they could have flown home from Kitty Hawk.

There was sadness in this madness, as well. Langley's reputation had been staked on the Aerodrome, and the two failures—on October 7 and December 8, 1903—were savagely covered by the press. He died heartbroken as a result of his failures, his sorrow made keen by the Wrights' success.

## THE FINAL PIECES TO THE PUZZLE

While Langley was striving toward ever more elaborate ephemera in his quest, the Wrights concentrated on the triple task of creating an engine of sufficient horsepower, yet light enough to permit flight; learning to make propellers efficient enough to absorb the engine's power and transform it into thrust; and devising a means of getting the power from the engine to the propellers.

The Wrights first wrote letters to more than a dozen automobile manufacturers, seeking a lightweight engine about 8 horsepower, which they believed would be adequate. But no automobile engines of this type were being built; on four wheels, ruggedness was more important than light weight, and there was simply nothing like what they wanted available. So, characteristically, they decided to build their own. They turned to Charles Taylor, the gifted mechanic who ran the bicycle shop in their absence, to execute the design.

He had almost no experience with engines, but he was a genius in a machine shop; and following the Wrights' instructions he built a four-cylinder engine, using only a lathe and a drill press. Taylor painstakingly drilled the crankshaft—the hardest part to finish and most critical part to balance—out of a steel billet. He finished it to rough shape with a hammer and chisel and smoothed it out on a lathe. Upon completion it balanced perfectly.

Taylor's approach was similarly direct with the rest of the engine. He bored out a cast block of aluminum to provide the cylinders, and he fashioned the pistons and rings out of cast iron. By today's standards, the accessories were primitive.

There was no carburetor; fuel dropped down a tube from the homemade gas tank to a chamber in the manifold and vaporized there into the cylinders. There were no spark plugs, the ignition being provided by a "make or break" system as simple and direct as a telegrapher's key.

And yet it ran well, with a throaty rumble not unlike that of a Model T Ford operating with a cutout exhaust. The engine developed almost 16 horsepower when it first started, turning up about 1,200 revolutions per minute. Power dropped off after the first fifteen seconds, however, to a steady 12 horsepower. By chance, this was just enough to permit the airplane to fly, even though the Wrights

J. T. C. Moore-Brabazon was the first man to fly in England; he joined the Flying Corps in 1914 and pioneered photographic reconnaissance.

**OVERLEAF:**
Progress doesn't always mean increased satisfaction, and the hang glider became for some a means of experiencing flight as the pioneers had.

had thought that 8 horsepower might suffice. In hindsight, Orville Wright later expressed some chagrin at the power output, saying that 12 horsepower was about half of what they should have obtained, given the engine's piston displacement and turning speed.

The propellers were a surprising stumbling block. Propellers had been used on ships for almost a century, and the Wrights expected to find an enormous body of knowledge about them. In fact, quantitative information was almost nonexistent; evidently, propellers were built by guess and by God in the shipyards. The problems confronting aircraft propellers were complex, having to do

with the speed at which the blades turned, the designed angle with which they met the air, and their shape. The first two factors could be expected to vary with changes in airspeed.

The Wrights' earlier experiments with a wide variety of airfoil shapes in the wind tunnel helped them immensely. They saw that the propeller was essentially a rotating wing and that the airfoil data they had derived from the tunnel could be applied to it. It was apparent to them that two larger propellers rotating slowly would be more efficient than a single smaller propeller rotating swiftly. Based on this, they created two propellers, one a mirror image of the other, for their 1903 experiments. They adapted a chain-type drive from their expe-

**When the ideas of hang glider and lightweight two-cylinder engine were joined, an entirely new air vehicle developed.**

rience in cycling. The transmission system, reminiscent of, but much stronger than, a bicycle drive, served to reduce the propellers' speed of rotation to about one-third the speed of the engine. It is a further illustration of their elegant thinking that they decided to have the propellers counterrotate, to offset any torque effect.

Difficult though their task was, they continued to test in the most scientific manner. They carved a solid propeller to their specifications, and tested it successfully. They next built two propellers—a left-hand and a right-hand—from three laminations of spruce and covered them in canvas. The propellers were extremely important, so much so that if they had been furnished by a collaborator (August

Herring, for instance, or Octave Chanute), the collaborator would certainly have been recognized ultimately as coequal with the Wrights in solving the problem of flight.

Yet there were no collaborators; the Wrights did it on their own.

This monumental work, so clearly superior to the work being done by all of the contemporaneous would-be practitioners of flight around the world, was all done in the single "off-season" between October 1902 and September 1903. The creation of the power train—engine, propellers, and transmission—was an incredible concatenation of leading edges. To have accomplished this and at the same time to have designed and prefabricated the

**The appeal of both the hang glider and the ultralight lies in the immediacy of flight. In this twin-engine Lazair, one literally becomes the wind.**

rest of the aircraft exceeds ordinary understanding.

The Wrights' capability as pilots was yet another aspect of their leading edge. They did not go to Kitty Hawk as Langley went to Quantico, intending to launch a machine they did not know how to fly. They could not, of course, have already experienced powered flight, but they understood that it would not be substantially different from gliding flight—the engine merely substituting for gravity as a power source. Their care in preparing themselves to fly gave them the best possible chance of preserving an aircraft that flew.

Still, much could have gone wrong. The Flyer was so extraordinarily sensitive that today a trained pilot could not step in and fly it without first going

through the same gliding/training process the Wrights did. As it happened, the first flight attempt, by Wilbur on December 14, ended with a minor crash when he overcontrolled. A very slight difference in wind velocity or in Wilbur's reaction could have ended in a crash that wrecked the Flyer beyond repair.

On December 17, when at last Orville made the first successful flight, many things could have gone wrong; and any one of them could have made the flight impossible. The engine might not have started, especially given the low-quality gasoline that was available. (Even today, routinely dependable engines manufactured to exacting standards for cars, planes, or lawnmowers sometimes simply choose not to start. That the hand-built, rough-hewn Wright engine started on demand is almost a miracle.) The propellers might have shattered or a propeller shaft might have cracked, as one did the previous September. The weather was bad enough that it would have kept a majority of today's Cessna pilots home in bed. Ice-covered ponds greeted them on the morning of Decem-

ber 17, and the wind whistled at 27 miles per hour. As wild as the wind was and as saillike as the 500-square-foot wings were, the gusts might have tumbled the airplane before the flights instead of after them (as happened).

Yet all of the slender margins were preserved, and the flights were made. The first covered 120 feet over the ground, equivalent to 450 feet in still air; the fourth and longest covered 852 feet over the ground, equivalent to almost a half mile in still air. It was the first time in history, as Orville put it, that a machine carrying a man had raised itself by its own power into the air in full flight, had sailed forward without reduction of speed, and had finally landed at a point as high as that from which it had started.

He might also have said that all of the leading edges came together in harmony. Certainly, the minimalist assemblage of intuitive and scientific deductions that underlay the Flyer were far ahead of anything in the world for the next five years. Even after the Wrights had inspired Europe with demonstrations in 1908 and 1909, their copiers took another two years to catch up. The fact that they ultimately surpassed the Wrights (by 1911) in no way detracts from the Wrights' incredible four-year accomplishment, which saw one serious scientific puzzle after another solved in a persistent systematic manner. After the Wrights, leading edges were never again developed for so many aspects of flying in so short a time. Moreover, later advances tended to be introduced gradually—sometimes coming together, but more often individually —and tended to have varying degrees of importance to the general problem of flight.

The Wrights gave the world twin legacies. The first was the basic shape and idea of their aircraft, and the second was their scientific method with its scrupulous attention to detail. Both were largely ignored by engineers in the United States, with a few exceptions such as Glenn Curtiss. In Europe, the first legacy was embraced with fervor, while the second was rejected for many years—until the progress built upon the first demanded the process to be derived from the second.

**LEFT:**
Once a workhorse trainer, now the sport plane of film stars, the de Havilland Tiger Moth flies on.

**RIGHT:**
The classic Great Lakes biplane, a product of the early 1930s, is periodically brought back into production to supply the need for an agile sport plane.

# 2. BLERIOT'S TRIUMPH:
## THE MONOPLANE

The Wright brothers functioned as a unit, two brains acting as one, two personalities bringing out the best in each other. Yet the attainment of the next clear-cut leading edge—the Bleriot monoplane—required a whole team of dedicated men.

Louis Bleriot is a perfect example of the human factor in the development of aviation. He won immortal fame on July 25, 1909, when he flew a brief 36½-minute flight of a mere 26 miles, a flight distinguished only and forever by the fact that it was from a tiny field near Calais, France, to New Foreland Meadow, near Dover, England. La Manche, which had thwarted Napoleon in the past and would frustrate Hitler in the future, had been breached. England was no longer an island, and a new aircraft company found itself soundly launched.

The aircraft was the Bleriot XI, the eleventh airplane Bleriot had participated in designing and the first to be notably successful as a flyer. The thirty-six-year-old, sharp-featured Bleriot reaped the fame of the flight, but his success depended as much on the team of about a dozen people who worked in his factory as on himself. One of them, Raymond Saulnier, would go on to found Morane-Saulnier and eventually become a more successful manufacturer than his mentor, designing aircraft of a similar controversial nature.

Team efforts would increasingly become char-

acteristic of aviation, and especially of innovation at the leading edge. Sadly, another characteristic of the phenomenon was that most of the team members who participated in the advance were deliberately or inadvertently forgotten, and their names are lost to history.

In the custom of the times, Bleriot had taught himself to fly in aircraft of his own design. It was fitting that an airman should learn to fly in cautious, intermittent steps in aircraft of his own design that developed equally intermittently: the flyer and the machine progressed together. In his career, Bleriot had no less than thirty-three crashes, and he walked away from all but the last.

While the Wrights had allowed the bicycle shop to generate a modest flow of funds to sustain their experiments, Bleriot drew deeply upon his own resources, derived from the automobile headlight firm he had founded and from his wife's dowry. Bleriot spent his time and fortune between 1905 and 1909 in a valiant, varied, and largely fruitless attempt to achieve flight. Like all pioneers of the period, he took ideas freely from everyone, putting combinations of these together that reflected several states of the art at once. His designs progressed not by leaps and bounds, but by fits and starts; Bleriot's I through V did not fly in any genuine sense, merely hopping off the ground before ending in a series of inelegant crashes. Bleriot VI

did rather better, flying for as much as 600 or 700 feet in distance, and 75 feet in height. It, too, crashed—perhaps fortunately, since it had the Langley tandem configuration and was probably an aeronautical dead end. The VII was a step in the monoplane direction, but it too was a failure. The Bleriot VIII was yet more successful, a precursor of the channel-crosser.

Ironically, as Bleriot expanded his activities, he seems to have spent less effort on what proved to be his greatest success, the Bleriot XI, than on any of the others. It was little more than an elaboration of the Bleriot VIII and did not offer the dramatic potential of his other developments at the

**Preparing his monoplane for the 1909 flight, Bleriot used the most ordinary support gear: ladders and sawhorses.**

time. Yet it was to prove a golden vehicle for Bleriot's reputation and personal finances, and it bore the brunt of the Bleriot companies' fortunes for almost five years.

Bleriot, like many of the pioneers of the time, was not himself a precise engineer; rather he had an instinctive flare for verbal description that (for a time) effectively removed the need for the sort of mathematical analysis given to the prosaic: bridges and Eiffel towers. This characteristic would ultimately destroy him and would temporarily—but for a crucial period—have a significant adverse effect on aviation.

The Bleriot XI was built simultaneously with the

much larger Bleriot X (which looked sort of like a Wright biplane with a pituitary problem) and the radical-appearing Bleriot IX. Neither the IX nor the X flew—to the advantage of the designer's continued good health, for Bleriot resolutely tested his own aircraft, deriving from the experience an intuitive sense of their problems and refining them in accordance with the knowledge he gained.

## THE FIRST INNOVATION

The monoplane formula of the Bleriot XI was an enormous departure from the Wright formula. A leading edge of vast though uncertain potential in

Bleriot's wife, Alice, shown here with Bleriot and Alfred Leblanc on her left, fully supported her husband's efforts both morally and financially.

1909, with its promise of both speed and danger, it is the standard configuration today. The Wrights had chosen the biplane configuration because they could create a strong, flexible structure using a Pratt truss. The monoplane required stout spars and excellent bracing to be strong enough to impart its advantages. It had less drag than a biplane, so for equal power it was faster. It had better visibility by far than the biplane, and it was less expensive to manufacture and transport. Monoplanes had disadvantages, too: they were often trickier to fly, and they almost always had a higher landing speed than a comparable biplane.

The Bleriot XI established itself as a leading

edge by conquering the channel, and thereby it found a magic formula for business success. Everyone wanted monoplanes—a Bleriot if possible, or if not, something very like it. The airplane, after the Wrights', was the most copied in the world, and Bleriot's factory had to expand to meet the burgeoning demand. Within two years, he employed over 150 engineers and had produced over 500 Bleriot XIs, in dozens of variations; production would exceed 800 by the time the advances of World War I rendered it obsolete.

What were the characteristics of this aircraft that brought Bleriot fame, wealth, and international status? The Bleriot XI was engagingly simple by today's standards, and we must compare it with earlier and later planes to gauge its impact. Its pri-

**The triumph of George Chavez's flight across the Alps from Switzerland to Italy was marred when—in classic Bleriot fashion—his wings collapsed and he crashed to his death.**

mary difference from other aircraft of the period was the monoplane wing, which quite unabashedly used the same wing-warping that the Wrights advocated. The wing was just over 25 feet long and had a chord (width) of 6.6 feet, for a wing area of 46.2 square feet.

The construction of the wing appeared robust, and the word "appeared" is used advisedly because within the general configuration lay ominous seeds of destruction. The fate which befell the Bleriot XI would stalk many other aircraft at the leading edge: brilliant innovation based on current knowledge often levered aircraft design into areas of the unknown, where previously unimagined dangers lurked.

The Bleriot wing consisted of two main spars

and four smaller spars, both surfaces covered with a rubberized fabric. The flying wires (those running from a post underneath the fuselage to the wings, which bore the loads induced in flight) were attached to each spar. Landing wires (which handled the loads on the ground) were guyed from the wing to a steel-tube pylon just forward of the pilot.

Thus the construction was somewhat primitive and pragmatic; static testing of the wing's strength in 1909 was even more so. No engineering handbooks were available for guidance, and testing consisted primarily of spreading sand over the wing in weights that were considered to be in excess of loads to be expected in flight. Unfortunately the testing was static, and flying is always dynamic.

Race courses were favored places to fly because of their open spaces—and rich patrons.

**OVERLEAF:**
Jan Olieslager—champion bicycle and motorcycle racer and six-victory WWI ace—taking off.

## ACROSS THE CHANNEL

For the early flights—and for the channel crossing on July 25, 1909—the Bleriot XI was powered by a three-cylinder Anzani engine which generated 30 horsepower at its full 1,600 revolution-per-minute speed, giving the type XI a maximum speed of about 40 mph.

This engine was adequate for Bleriot's great task; when he took off from France, he immediately throttled down to save the engine and flew at 250 feet above the sea. It was an uncomfortable altitude for an uncomfortable flight. He had taken off reluctantly under obvious stress, concerned about the weather and the hazards of a descent into the channel. Once airborne, he was

so preoccupied that he ignored the compass on board and steered almost by instinct. Undoubtedly the fact that his foot was still aching from having been burned twice in previous accidents—he had hobbled to the airplane on crutches—added to his misery. Finally, the white cliffs of Dover came into view, and a colleague, sent ahead, was standing in a breach in the cliffs waving a huge tricolor, urging him on to land.

A wave of enthusiasm surged around Bleriot—one not to be matched in response to a flyer until Lindbergh landed at Le Bourget in 1927. Bleriot was the hero of the hour; English editorial writers pointed out ponderously that "England was no longer an island," while French editorial writers proclaimed that Bleriot had regained from the Wright brothers the glory of flight for France.

## MASS PRODUCTION AND DESTRUCTION

Orders for type XIs poured in; by September, more than 100 were on order, and it was necessary to expand the factory. With the orders came demands to increase the performance, for the entire world of aviation was trying to eclipse what Bleriot had done, and the new aircraft all promised greater performance.

Bleriot responded with an entire line of aircraft—two-seaters, parasols, floatplanes, racers, and trainers, for both civil and military use—each equipped with whatever engine the purchaser wanted to use. But therein lay the rub: new engines had appeared on the scene that were far more potent than the erratic, ill-balanced, 25-horsepower, three-cylinder Anzani that had carried him across the channel. The new series of Gnome rotary engines (leading edges themselves, and discussed in chapter 6), rated at 50, 70, 80, or even 140 horsepower, were installed in Bleriots of various types; and fitted to these powerful, dependable —relatively—engines were Chauviere propellers, vastly more efficient than anything Bleriot had employed before. (In the past, he had used propellers that looked like four spatulas set at right angles to each other. The Chauvieres approached the

**RIGHT:**
A Bleriot VIII at Issy-les-Molineaux, a field still to be seen in Paris. Note the primitive propeller.
**OPPOSITE:**
The Bleriot formula was soon copied by other manufacturers and often improved upon—as in this Morane-Saulnier, an early member of a distinguished line of combat aircraft.

Wright brothers' propellers in efficiency.)

Speeds rose to as much as 70 miles per hour within a year of the channel flight, and to 80 mph within a year after that. Accidents began to happen, too, killing the best of the Bleriot pilots and sharing a chillingly familiar characteristic: the collapse of the wings.

The first occurred when Leon Delagrange, a charming, immensely popular man, died on January 4, 1910, when his aircraft's left wing folded. It happened again in April to Hubert Leblon, who had investigated Delagrange's death and decided that the accident occurred as a result of one of the flying wire's breaking. Other fatal crashes followed, and the monoplane in general and the Bleriot in particular obtained an infamous reputation. Louis Bleriot met the challenge in his customary straightforward fashion. He investigated the causes and said, in effect, that the reasons for the accidents were unknown but that they must in part be due to deficient strength in the aircrafts' bracing. Testing and engineering were not sophisticated, so Bleriot's empirical method of simply increasing the strength of the spars and bracing wires was accepted.

The strengthening measures were inadequate. More accidents occurred, ultimately leading to military grounding of the aircraft in both France and England.

The real problem did not lend itself to analysis until many years later, when engineering had

become more sophisticated. In layman's terms, the Bleriots were simply overstressed. More powerful engines permitted speeds above 50 miles per hour; under certain conditions which included the higher speed, the effect of g (induced "gravitational") forces, and the manner of control input by the pilot, the air loads imposed by these speeds had the effect of twisting the tips of the wings upwards. As the wings twisted, the air load progressively increased, leading inevitably to torsional failure.

The failure mode was identified with monoplanes surprisingly quickly, and the type was immediately suspect. In 1912, after a series of accidents, England placed a ban on all monoplanes.

The imminence of war and the fact that a large proportion of the small English air fleet consisted of monoplanes caused the restriction to be rescinded, but a pall was undoubtedly cast upon the configuration, not only in England, but around the world. The margin of performance between a monoplane and biplane was so slight (in practical terms it did not matter if the top speed was 50 or 60 mph) that buying the more rugged biplanes was the sensible course.

Nonetheless, the Bleriot XI and its descendants had a profound effect on the industry; and just as Bleriot had copied essential features of other designs, other manufacturers copied (and improved on) the Bleriot.

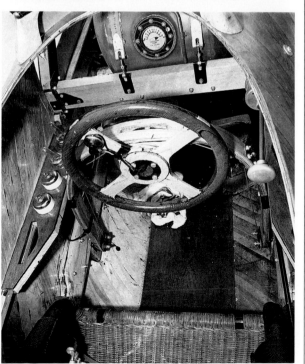

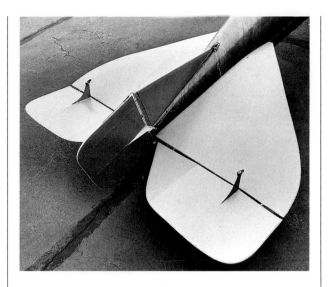

## THE SINCEREST FLATTERY

A majestic progression of design ideas advanced from Bleriot to a whole host of manufacturers. As so often happens in the industry, the basic design was carried from one company to another in the mind of a designer. The number of aircraft manufacturers was small, and it was natural that the emerging Nieuport company would create aircraft not only reminiscent of the Bleriots, but superior to them. Saulnier left Bleriot to join the Morane brothers at Morane-Saulnier, and there created a line of aircraft with obvious Bleriot qualities. The circle spun still wider: Anthony Fokker admired the Morane-Saulnier so much that he acquired a damaged example on which he based a whole line of Fokker monoplanes that led directly to the Eindecker and the "Fokker Scourge" of 1915.

## BREAKING THE 100 MPH BARRIER

A young French engineer, Louis Bechereau, had studied both Bleriot and Nieuport designs before joining the small firm of Armand Deperdussin—the Société pour les Appareils Deperdussin—in 1911. There he created the most advanced airplane of its day, the Deperdussin racer, using (in a radical departure from conventional practice) the monocoque construction of the Swiss Ruchonnet. Three

layers of thin tulipwood veneer were laid around a cigar-shaped mold and glued together to form a lightweight, strong, well-streamlined fuselage. Bechereau then boldly chose an entirely new powerplant. The rotary engines that had overpowered the Bleriots had been further advanced by combining two engines into one. Bechereau incorporated one of these 140-horsepower twin-row Gnome rotary engines into his design, combining it with a propeller with a large spinner that was streamlined neatly into the nose of the racer.

A final element marking Bechereau's work was his elaborate attention to detail. He faired in the wheels with covers, put a streamlined coaming behind the pilot's head, and in general minimized the effect of small drag-producing items by attending to them with a degree of care that had not been seen before.

The results were sensational—and all the more dramatic because preflight critics had labeled the plane as too dangerous to fly. It was not too dangerous for Jules Vedrines, a pilot of notorious temper who flew the airplane to a series of records, achieving more than 106 mph by July 1912.

## AT THE EDGE OF WAR

In nine years, the speed and design of aircraft had moved from the Wrights' 31-mph saunter to speeds exceeding the century mark, and with it had moved the military conception of aircraft as an instrument of war.

The Deperdussin was revolutionary in concept. Its designer, Louis Bechereau, went on to produce France's finest fighters during WWI.
**OPPOSITE:**
Here two classically beautiful monoplanes—the Antoinette (left) and Hanriot (right)—race along at speeds approaching 70 mph (*top*). The Deperdussins were the fastest things in the air. Brakes were of no consequence—no one worried about being able to slow down (*bottom*).

The early pioneers of aeronautics had established the basis for another great leap forward: the creation of the fighting aircraft of World War I. Fokker, a Dutch national, parlayed his Morane-Saulnier–inspired designs into a series of handsome fighters culminating in the Fokker D VII and Fokker D VIII, which were certainly the best German fighters of the war and perhaps the best on either side.

It would be discovered in 1913 that Armand Deperdussin's gloriously open-handed career in aviation was founded upon 28 million francs of other people's money, illegally obtained; and he went to jail. Bleriot reentered the picture by reorganizing the company, keeping Bechereau as chief designer, and retaining the initials of the original Société pour les Appareils Deperdussin by calling it the Société pour Aviation et ses Dérivés. The company, whose acronym was rendered as Spad, manufactured the best of the French fighters of the war (these were used by the American flying service, as well). By the end of the war, Bleriot-owned factories were producing eighteen aircraft per day.

Thus the Bleriot XI (derivatives of which were used in the early days of the war) was not only the first airplane to cross the channel, but in many ways the progenitor of a whole series of types put into combat on both sides for the next four years. When the war ended, the Bleriot XI and similar types had become museum pieces that were no longer useful even for instruction, but in their descendants was writ large the destiny of air warfare.

# 3. THE LEADING EDGE

## AS SWORD AND SHIELD

In the fascinating history game of "what if," there is a seldom-played minor variation: "What if the tiny British Royal Flying Corps had not monitored the German army's progress through France in August 1914?" This seemingly innocuous "what if" inquires into the effects of the observations of the two dozen frail British Bleriots and Farmans that were deployed to France on August 14 and flew from apple orchards and wheat fields during that fateful month.

Yet the consequences of the British air monitoring are with us to this day. Consider: the Germans had launched the von Schlieffen plan, according to which a monstrous southwesterly sweep through Belgium was to cut between the French and British armies and swing around Paris, defeating France in six weeks, bringing the British to the negotiating table, and leaving the German railway system ample time to transport the field armies east to defeat Russia. If that had happened, we can say immediately that Hitler would not have risen to power, there would have been no Soviet Russia, and the entire course of world history would have been altered. Indeed, the alteration might have been for the better, if we imagine a peaceful Germany policing the European continent for the next sixty years (much as England had kept the peace at sea during the previous century) and an Anglo-American alliance being forged to contain it.

The primary reason this did not happen lies at the heart of British aviation and, in a way, at the heart of the English character.

## THE FIRST OF THE FEW

In 1910, a tiny aristocracy of aircraft builders and flyers had begun to gather at Larkhill on Salisbury Plain, not far from Stonehenge, to make periodic attempts to coax a wheezing series of aircraft into the air. They learned to fly at an average height of 30 feet above the ground and at speeds of less than 50 miles an hour. The transition time from absolute novice to solo was measured not in hours, but in minutes: a clever student, mindful that all turns must be made to the left to gain assistance from the gyroscopic action of the whirling Gnome engines, could become an instructor with three or four hours of total flying time under his belt—free to pass on his knowledge (or lack of it) to the next willing, hapless victim of the urge to fly. The comic aspects of this situation were more than offset by repeated tragic crashes. Yet this was the foundation of the legendary Royal Flying Corps, which became (on April 1, 1918) the Royal Air Force.

The Royal Flying Corps was a rarity in the tradition-steeped British armed forces: it was a haven for eccentrics, mavericks, and the wildly adventurous, for young men who could not see the

wisdom in a posting to the cavalry or the infantry. In practical terms, the early pilots were aerial remittance men, wearing the uniform of their home regiments, but forfeiting their careers to indulge their wild desire to fly.

Yet the traditional British army maneuvers of 1912 and 1913 made use of the aircraft as a reconnaissance vehicle, and in doing so prepared its flyers for duties of immeasurable consequence in 1914. In those choreographed grapplings between four divisions of Sir Douglas Haig and a smaller force under General Grierson, it was found that the observation aircraft, primitive though they were, could furnish information at least twenty-four hours in advance of that provided by the traditional

A Nieuport 28 in an undignified—but all-too-familiar—posture. The French had rejected this plane for front-line service, but they were issued to American flyers when they arrived as the only available airplane.

cavalry scouts. These practice wars not only prepared the handful of pilots and observers, they conditioned the minds of the top British generals (no easy task) to accept information delivered by air. They also caused plans to be made for a move to France in case of war, one of the very few inter-army arrangements that were worked out in minute detail before the war.

## THE AIRPLANE GOES TO WAR

When war came, as everyone both knew and denied must happen, the ragtag collection of observation aircraft crossed the English Channel in a mass reversal of Bleriot's flight—each pilot laden

with a revolver, field glasses, spare goggles, a water bottle, a small stove, biscuits, cold meat, a piece of chocolate, and soup cubes. The four squadrons (twenty-four aircraft) that left for France on August 13 consisted of a mixed bag of Bleriots, Henri Farmans, Avros, and B.E. 8s, all fragile and temperamental.

The rest of the corps moved by ship, carrying with them a wide variety of ammunition. The bomb lorry of No. 5 Squadron was painted brilliant scarlet, and carried the slogan, ''The World's Greatest Appetizer'' on its side. Thus Worcestershire sauce went to war.

They began flying the familiar observation flights they had practiced on maneuvers; and on Au-

The salvage and repair efforts during WWI reached a surprising peak of efficiency; and this Nieuport 28—or at least parts of it—was soon returned to service.

gust 23, Captain L.E.O. Charlton and Lieutenant V.H.N. Wadham of No. 3 Squadron observed von Kluck's army wheeling to the right of the British, seeking to envelop Paris.

Given the intransigent stupidity of the Allied leaders, who sent blue- and red-uniformed innocents in measured march against the German machine guns time and again, who believed in the attack until no more bodies remained to attack with, and who stayed aloof in chateaued refinement from the degrading, degenerate misery of the trenches throughout the war, it is incredible that both the English and French high commands believed the aerial reconnaissance reports. Moreover, they acted upon them, redeploying troops to slow the Germans down and eventually forcing them into the decisive Battle of the Marne.

From this noncombatant beginning, the derivatives of the Wright and Bleriot aircraft were soon adapted to perform virtually every form of aerial warfare known today, except nuclear attack. The British initiated the practice of strategic bombardment with their preemptive raids on German zeppelin hangars; the Germans originated aerial terror bombing with a few hand-held bombs lobbed on Dover; both sides employed airplanes for aerial reconnaissance, aerial combat, air-to-ground attacks, and even psychological warfare. By 1915, aerial observation and the registration of artillery attacks had become so common that the entire battle of Neuve Chapelle, begun on March 10, was based solely on aerial reconnaissance photos.

The artillery shooters and reconnaissance flights quickly attained such importance that special aircraft types—the earliest fighter planes—had to be deployed to counter them. Ultimately, the front that ran from Switzerland to the sea was backed on each side by hundreds of aerodromes, backed in turn by entire industries newly created to supply the proliferating varieties of aircraft.

## Newspaper Heroes

The war in the air was romanticized by the press at the time, as it has been by all forms of media since. There was a good reason for this: the conflict in the air was recordable for the general public, since it consisted of individuals engaging in single combat in an atmosphere of "glamorous" adventure. In contrast, the war on the ground simply could not be truthfully recorded; even apart from the problem of pervasive military censorship, the hideous slaughters at the Somme, Ypres, and Verdun could not be comprehended and described for public consumption. The newspapers had to have heroes, and they created them in the air.

Yet the fighting in the air was fraught with danger and horror. The pilots were usually not yet twenty-one and sometimes had as few as six or seven hours of flying time before being sent on

**OPPOSITE:**
Some aircraft seize the imagination and never let go. Such was the Fokker triplane.

**FOLLOWING PAGES:**
*Page 56:* **Significant in this Fokker triplane is the presence of the forward-firing machine guns (upper left).**
*Page 57:* **The Fokker D VII is considered by many to be the best fighter plane of WWI.**
*Page 58:* **The D VIII was nicknamed the "Flying Razor" by British pilots because its small frontal area was so difficult to see when attacking out of the sun.**
*Page 59:* **Modern pilots have nothing but admiration for the young Germans who flew the triplane, for it is a tricky plane to fly and even trickier to land.**

combat missions in aircraft they had never flown before. Aircraft themselves were still made largely on an intuitive basis, subject to great hazards of design and manufacture, and of course extremely vulnerable to machine gun fire. To a fuzzy-faced teenager, it was not glamorous to find that their mounts burned easily and fiercely, the 100-mph wind whipping an exploding gasoline tank into an inferno instantaneously.

For most of the war, too, no parachutes were issued, despite their being well developed, because commanders felt that wearing a parachute might encourage a pilot to leave a damaged aircraft prematurely. Late in the war (and fearfully short of pilots), the Germans—who were almost always fighting over their own lines—began to allow use of the device.

The cycle of war that began as the Germans slashed through northern France and then evolved into the stalled battle lines of trench warfare forced refinements in observation, bombing, and pursuit aircraft. Intensive research in aircraft design was conducted on both sides, and the French and English leading edges were soon countered by German ones. Thus the British B.E. 2c was overwhelmed by the Fokker Eindecker, a Bleriot/Morane descendant that became known as the "Fokker Scourge"; the Eindecker was in turn bested by the de Havilland D.H. 2 and the Nieuport 11; and these were driven from the skies by the sleek and deadly (both to enemy pilots and to their own, for they had an inherent structural weakness) Albatroses, upon which von Richthofen built his fame.

This seesaw process continues to the present day. It was illustrated in World War II when Messerschmitts met their equal in the Spitfires, in Korea when the MiG-15 fought the F-86, in Vietnam when later MiGs fought the McDonnell Douglas F-4, and today when F-14s and F-15s are matched by MiGs and Sukhois of comparable performance.

## The Air Weapon Develops

In World War I, the military situation produced different lines of development in the air fleets of

the Allies and those of the Central powers. The Allies had control of the seas, and resources from all over the world—particularly from America—poured into England and France. Some items were in short supply, but plenty of the essential materials (wood, copper, brass, steel, rubber, and petroleum products) required to create the finest aircraft were always available. Germany, denied by nature and the blockade any excess of these necessary products, had to improvise both its aerial tactics and the aircraft it selected for manufacture. So poor were the Germans that their aircraft were often fitted with wooden wheels instead of rubber tires when they were not actually going to fly, and Allied crashes were systematically scavenged for engines, brass, copper, and other scarce materials.

The Allies used myriad designs—indeed, too many for their own good. France alone produced forty-three different kinds of engines and scores of different types of aircraft. The same pell-mell eclecticism was practiced in England, although it was recognized as inimical to large-scale production and to maintenance in the field, and although attempts were made periodically to regulate the process.

The British were especially profligate. They expanded their air arm from the crucial few that came over in 1914 to tens of thousands of aircraft. These were sent out with their young pilots to fight aggressively all over the world. The French were not so aggressive (nor so immoral in their deployment of youth), while the Germans quite sensibly adopted a defensive strategy; they fought over their own lines and defended only when they deemed it important.

## THE BEST OF THE BREED

By 1918, two Allied fighters, the Spad and the S.E. 5a, reigned supreme. Pilots perhaps favored the S.E. 5a, since the Spad was flawed by inherent problems with its Hispano-Suiza engine; yet both epitomized the final development of the Wright/Bleriot wood-and-wire formulas. The Germans, on the other hand, fielded the utilitarian-looking Fokker

Hugo Junkers (*top*), and Reinhold Platz.
**BELOW:**
Anthony Fokker.

D VII, which foreshadowed the aerodynamic leading edges of the future as surely as Spad and S.E. 5a symbolized the past. How did they compare?

The Spad was designed by Louis Bechereau, of Deperdussin fame, with his usual attention to detail and his remarkable ability to create strong, complex structures that were nonetheless easy to mass-produce.

Bechereau had learned of the new and reportedly wonderful 150-horsepower Hispano-Suiza V-8 engine that had been designed by Marc Brikgit, a Swiss engineer working in Spain. The Hispano-Suiza engine was extremely modern in concept, with a favorable power-to-weight ratio that made it ideal for use in a fighter, and Bechereau created the Spad VII around it. The Spad VII was a rugged biplane fighter that went to the front in late 1916, just in time to offset the ascendancy of the German Albatros (then appearing in great numbers).

Bechereau, like all designers, sought more power, and the Hispano-Suiza engine was upgraded to 220- and 235-horsepower versions—adjustments that required special gearing to reduce propeller revolutions to a reasonable level. The bigger engine also permitted a slightly more sophisticated Spad to be developed, the XIII.

The Spad XIII's engine and performance represented the leading edge of the French aviation industry; at the same time, the aircraft concentrated a number of otherwise obsolete techniques and ideas. The box-section fuselage, for example, followed tradition in being composed of wooden members and metal joint fittings, the whole being held together with piano wire bracing. The wings featured hollow box-section spars made in short sections and united by scarf joints that were bandaged with linen and doped. The French lacked spruce in sufficient lengths, so the spars had to be cobbled together.

Yet so intricate was the construction and so redundant was the bracing that the Spad XIII made a tough and rugged fighter. It was, however, betrayed by its uprated engine, which was prone to failure from a number of sources. As many as two-thirds of the Spad XIIIs at the front were unavailable for service at a given time because of engine maintenance problems.

Its classic antagonist, the Fokker D VII, came from the distant but familial lineage of the Bleriot, passing through Morane-Saulnier's interpretation and then through a development process unique in history. Anthony Fokker was twenty-four years old when World War I broke out. An entrepreneur, he reportedly had offered his services to the Allies and had been refused before setting up business in Germany, where he created some of the most formidable (and some of the most ridiculous) warplanes of the period. Fokker was a brilliant pilot and an intuitive designer, but he relied heavily on the engineering expertise of others. By 1917, this advisory role had devolved primarily upon the shy and retiring Reinhold Platz, who had been his chief welding engineer. Fokker sequestered Platz in Schwerin, denying him access to available engineering reports on other current Allied and German aircraft, and then called forth one brilliant design after another from him merely by stipulating the requirements and the desired configuration.

Platz worked well in this enforced isolation, and perhaps Fokker was wise to nurture his genius and prevent its cross-pollination with other, lesser ideas. Platz believed in simplicity, lightness, and ease of manufacture, and he achieved these brilliantly in the Fokker D VII.

The D VII featured a welded steel-tube fuselage, a Fokker trademark that was adopted all around the world after the war, and thick biplane wings. Platz saw—before the experts running the wind tunnels did—that a designer could make thick wings that had a lift-over-drag ratio superior to the ratio attainable using the thin wings of the time. Thick wings had great advantages: they could be made stronger and lighter for their size than thin wings, and they could obviate the need for all or most of the strut-and-wire bracing that created drag on thin-wing airplanes.

By 1918, Germany was pitifully handicapped by a shortage of essential manufacturing materials, precluding the development of more powerful engines. As a result, the matchless D VII airframe

**RIGHT:**
Relatively slow but extremely maneuverable, the Fokker Dr I was effective in the defense mode used by the German Air Force (*top*). The Fokker D VII was crude in appearance, with its flat car-radiator nose and slab-sided fuselage. But it "made good pilots out of bad ones, and aces out of good ones" (*center*). Cockpits in WWI planes—as in this Spad XIII—were tiny, cramped, and austere (*bottom*).

was fitted with an obsolescent 160-horsepower Mercedes powerplant and was sent out to combat the much more powerful Spads and S.E. 5as. Nonetheless, the clean design and the effective thick wings enabled the D VII to prove itself the best fighter of the war. Late in 1918, the D VII was equipped with a BMW engine of 185 horsepower, making it an even more superior aircraft. The D VII was said to make good pilots out of poor ones, and aces out of good ones.

So important was it that the D VII alone of the myriad weapons developed in World War I was singled out by name in the Treaty of Versailles. Article IV of that treaty detailed the material to be turned over by the Germans and stated "especially all machines of D VII type."

Perhaps even more important than its wartime role, however, was the influence it had on the development of fighters elsewhere. In Britain, Russia, and especially the United States, fighters followed the Fokker/Platz steel-tube and thick-wing formula for the next ten years. The Spad formula was abandoned, a relic of its times.

Curiously, when the Spad and the Fokker crossed paths in the sky, their performances were so nearly equal that the outcome almost invariably depended upon the skill of the pilots. Yet the Spad was a product of the past, while the Fokker was the forecast of the future.

## THE FIRST MASS-PRODUCED ALL-METAL AIRCRAFT

Among the other contemporaneous forecasts of the future was the all-metal aircraft of Germany's Hugo Junkers, a man of remarkable versatility and vision. In 1910, Junkers patented a fully cantilever flying wing of enormous size, and shortly thereafter he began development of a diesel aero engine.

In 1915, he designed (in company with two other engineers, Mader and Reuter) the Junkers J-1, an all-steel, cantilever low-wing monoplane. The most modern structural methods were used, with the center section of the fuselage and the wing's center section built as an integral unit. The

**RIGHT:**
The Nieuport 28 was powered by a 160 horsepower Gnome rotary engine.

**OVERLEAF:**
The Curtiss JN-4 Jenny was America's principal contribution to the war effort, training thousands of U.S., Canadian, and British pilots. Slow and underpowered, it was forgiving enough to let youngsters solo after about a dozen hours of training.

Captain Edward Vernon Rickenbacker—America's ''Ace of Aces''—in the Nieuport 28.
**LEFT:**
The 1917 Curtiss S-3 triplane.

entire aircraft was covered in thin plates of sheet steel, welded to the basic tubular structure. The airplane was extraordinarily drag-free for its time, and it had a higher top speed (106 mph) than any of its observation biplane contemporaries. On the other hand, it was also far heavier, which impaired its maneuverability and climbing capabilities. Like most unusual aircraft, it acquired a variety of nicknames, ranging from "Tin Donkey" to "Flying Urinal"—a sobriquet derived from rudimentary comfort stations then found in city streets throughout Germany.

Junkers continued to elaborate on the basic formula, one design following another and all looking strikingly modern. In this first serious development of all-metal military aircraft, many engineering disciplines were as yet undeveloped, however—particularly in the area of weight control. The Junkers engineers' enthusiasm for the use of steel and their concern to have the monoplane strong enough led them to turn out aircraft that were almost invariably much heavier than planned, with a consequent detriment to performance.

Junkers persisted and eventually brought forth a series of fighter and observation aircraft that were far in advance of their time. Perhaps the best of the lot were the J-9s and J-10s, which only appeared during the last few months of fighting on the Western front and made little difference there, but which served with distinction in the series of border conflicts that Germany encountered in the east in 1919. One remarkable aspect of the planes was their invulnerability to weather: a Junkers could sit outside unharmed for months in weather that would have destroyed any conventionally built aircraft.

The Junkers monoplane formula was deviated from only once in World War I, and then with great success. The Germans had developed a system of "contact patrols," in which heavily armed attack aircraft operated in close cooperation with infantry attacks. For this purpose, Junkers designed the extraordinary J-4, a great all-metal biplane that was heavily armored and tremendously effective in its role. More than 200 were built and only

The MiG 21, in many ways the last of the classic "dog-fight" fighters that began with the Fokker monoplanes in 1915. In this supreme denouement of fighter tactics the hunter is above, the hunted below. And, excerpting Baron von Richtofen, "all else is rubbish."

1 was shot down, although they engaged in the most hazardous duty of the war.

The significance of the Junkers contribution was not understood for almost twenty years, when all-metal warplanes became the world standard. This delay occurred even though the idea was greatly reinforced by the conversion (after the armistice) of some J-10s to the first in a long line of Junkers passenger aircraft—a line that later included the famous six-passenger F-13 transport, the "Tante Ju" or Ju 52-3/m (which was Germany's primary civil and military transport), and the gigantic six-engine Ju 390 (which flew within 12 miles of New York City in early 1944).

Junkers led the way with metal technology, and other manufacturers followed. Notable among those were Dornier, with its sensational flying boats and its monstrous DO-X, and Rohrbach, which contributed substantially to all-metal design. But the world was not quite ready for all-metal aircraft as the standard, and for the next ten years the aircraft built were not greatly different from the ones with which Germany ended the war.

## A REDIVISION OF WARS

Progress often differs from diplomacy in its division of history. In levels of performance, tactics, and strategy, the first war in the air lasted not from 1914 to 1918, but rather from 1914 to 1943; although performance was improved over the peacetime years, the difference in technology and conception between 1918 and 1940 warplanes was not vast. It might be said that the second war in the air lasted from 1943 thorugh 1945, when aircraft were employed in a way that was fundamentally new in intensity, concentration, and effect. Finally (and regrettably), the third war in the air (which continues to this day) began with the awful combination of the B-29 and the atomic bomb. After World War I, the changes forecast by Bleriot, Fokker, and others emerged in a remarkable series of designs incorporating leading edges of varying degrees of importance and characterized by ever-sleeker shapes.

**LEFT:**
The Messerschmitt Bf 109 occupies a role in people's minds as the great German fighter of WWII—as the Fokker triplane does for WWI (*top*). Initially regarded as a failure, the Messerschmitt Bf 110 became a successful night fighter (*bottom*).

**RIGHT:**
German paratroops dominated headlines during the early days of WWII. Here they leap from a Junkers Ju 52.

**FOLLOWING PAGES:**
*Pages 70–71:* One wonders what a Rickenbacker could have done during WWI with a squadron of McDonnell Douglas F-15 Eagles, the Mach 2 front-line fighter of the USAF today.

*Page 71:* The mission of the bomber has not changed since WWI: bombs on target. Here the General Dynamics FB-111 used in the 1986 raids on Libya is refueled in flight.

*Pages 72–73:* The techniques of WWI persisted through 1943; the methods of WWII were made obsolete by nuclear weapons in 1945. Today's aircraft carriers pack on a single flight deck more firepower than was expended in the whole of both wars.

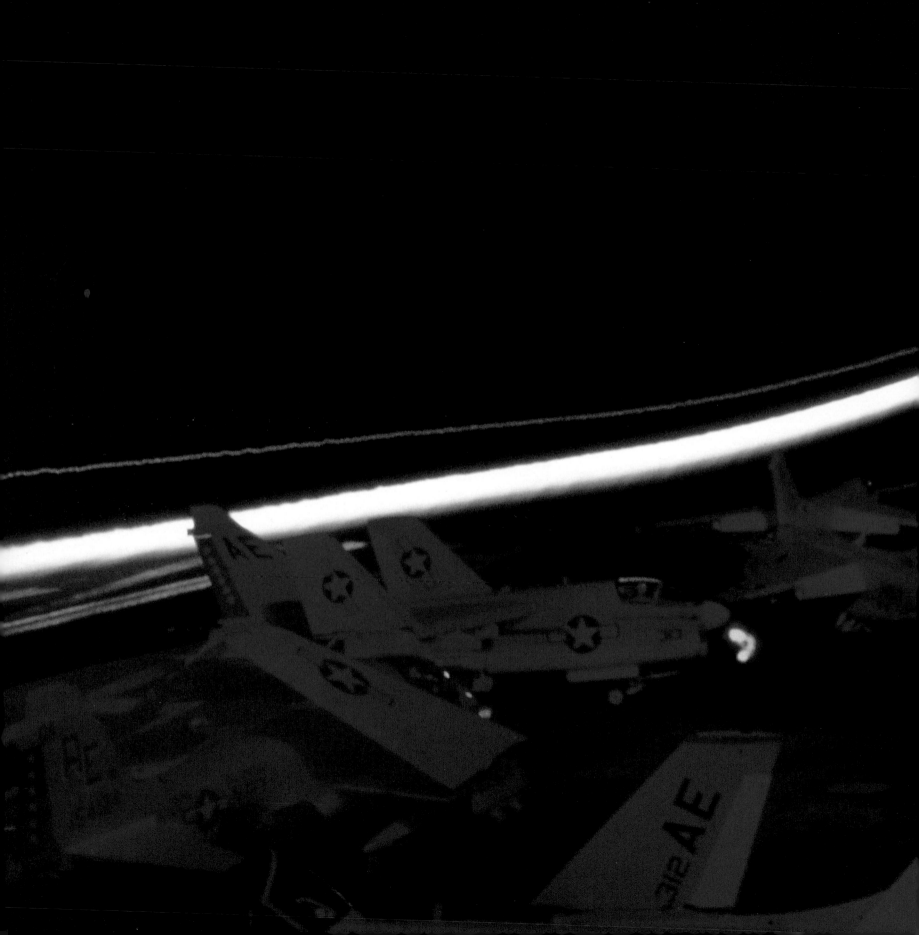

# 4. THE SHAPE

## OF THINGS TO COME

The Bleriot had established the monoplane as a leading edge until structural flaws inherent in the designers' pragmatic approach brought accidents that restored the biplane to favor. After the war, however, a number of monoplane designs reappeared, the most advanced of which was clearly the 1922 Verville Sperry R-3 racer. This clean, low-wing aircraft with retractable landing gear exhibited lines that would not have been out of place at the beginning of World War II. It is a classic example of a leading edge that did not realize its potential because of extraneous events.

The sleek little R-3, easily ten years ahead of its time, might have revolutionized aviation and established the United States as the world's leading air power during the late 1920s if a rational test program had been established to refine it. Unfortunately, it fell victim to a bureaucracy that was vulnerable to corporate pressure and to the then-common overreliance on test pilots' reactions.

### BLUNTING THE RACER'S EDGE

Until the jet engine became relatively well developed, advances in airframe design had always been slowed by limits in the amount of horsepower available. And since engines took longer to design than airframes, airframe designers always railed at their powerplant counterparts to hurry up.

Alfred Victor Verville.
**OPPOSITE:**
The Verville Sperry R-3—with its cantilever low wing, retractable landing gear, and smooth lines—was easily ten years ahead of its time (*top*). The R-3 was cleaned up with engine covers and wing skin radiators, enabling Harry Miller to win the 1924 Pulitzer race (*bottom*).

In a similar way, the matching of engine and airframe outstripped the development of instruments, land facilities for airports, air traffic control methods, and safety standards. These hazards, compounded by inadequate testing and unsophisticated test pilots, quite naturally stifled the progress of many new ideas.

The most human of these obstacles was the presumption that piloting constituted a better guide than science in the creation and analysis of new aircraft. The press attributed almost superhuman qualities to pilots, and even professionals rarely examined a pilot's qualifications closely. Test pilots in particular received this near-idolatrous treatment, and aviation suffered because of it.

At military testing fields around the world—McCook, Langley, Philadelphia, Farnborough, Villacoublay—the test pilots' position of authority corresponded to that of the original astronauts some four decades later. (The two groups also present striking similarities in personality, physical condition, motivation, determination, and ego.) They were the supreme arbiters of test flying, their word was law, and they could make or break an airplane with a single evaluation. Unfortunately, they were relatively inexperienced and they regarded the task of test flying as considerably less difficult than it really was.

Most test pilots were excessively concerned

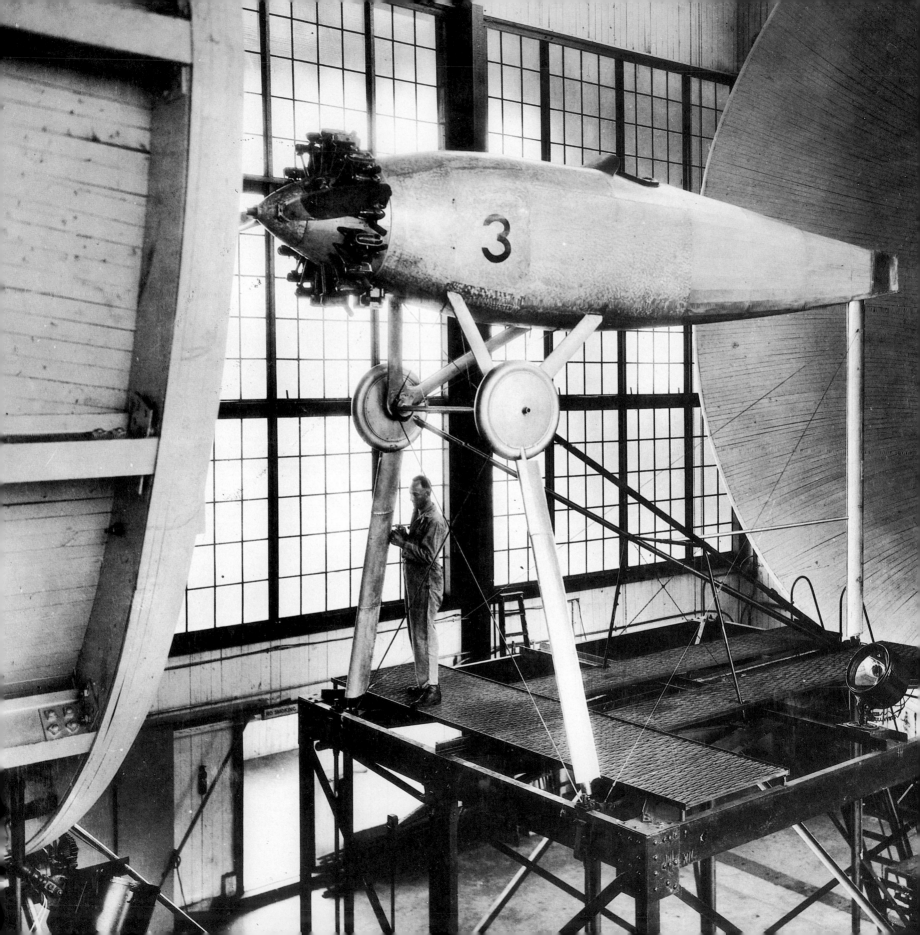

**LEFT:**
Wind tunnels played an increasingly important role in the development of aircraft. Early calculations were plagued by errors in scaling up results when models were used in smaller tunnels. Far better results were obtained when full-size aircraft could be tested under controlled conditions.

with the test aircraft's immediate combat potential. If it did not handle well, that was enough; test pilots rarely suggested that the long process of tweaking be undertaken to improve its characteristics. Thus, only the practice and not the science of test flying had been created.

On more than one occasion, the work of Alfred Verville fell victim to this condition. One of the most beloved engineers in aviation, Verville was a man of sweet, almost saintly disposition, and his ideas were far in advance of his contemporaries'. In 1922, he responded eagerly to a challenge by the famous and controversial General Billy Mitchell to create "tomorrow's airplane today." Mitchell's intention was to create a superfast airplane, win the prestigious Pulitzer Trophy with it, and then have the plane turned into a fighter with the addition of guns and self-sealing gasoline tanks. In this he anticipated the process of the subsequent Supermarine S.6, which was the progenitor of the immortal Spitfire of World War II. Verville designed his aircraft, the R-3, at McCook Field near Dayton, Ohio, and three were built in the factory of Lawrence Sperry at a cost of $25,000 each, sans engine.

The R-3 combined the Fokker-style steel-tube fuselage construction and thick cantilever wings with streamlining of the fuselage and introduction of retractable landing gear. The R-3 exactly presaged the fighters of the late 1930s, lacking only an enclosed cockpit.

Verville designed the R-3 to work with the most modern powerplant available: the Curtiss D-12 engine and the Curtiss-Reed metal propeller combination. Streamlining was improved by selecting the thin brass radiators Curtiss had developed for wrapping the sleek wings of its racers; this avoided the barn-door drag of standard types.

But Curtiss—at this time the most powerful manufacturer in American aviation—persuaded the Air Service bureaucracy to refuse Verville these three crucial items. This insistence on patent rights might have been understandable in a purely commercial situation, but the government exercised substantial control over aircraft manufacturers and could have insisted that the engine, propeller, and radiator patents be released for use with the R-3. Curtiss's influence prevailed, however, and a cranky, vibrating Wright Hisso engine with a great club of a wooden propeller was substituted on the R-3 for the potent Curtiss combination. Likewise, the sleek, airfoil-adapted skin radiators were replaced by Lamblin "lobster pot" radiators, attached to the sides.

The results were predictable: instead of running away with the Pulitzer race, the best any of the three Vervilles could do was to place fifth. The winner was a Curtiss R-6 biplane equipped with the very items Verville had planned on using.

The debacle of the R-3 was not exclusively the fault of Curtiss's proprietary penchant. A less obvious but more important problem was the total disinclination of aircraft manufacturers (including Sperry) to test aircraft properly. As late as the mid-1930s constructors would routinely roll out a racer the day before a race, expecting the genius of the pilot somehow to overcome all of the problems that arose.

The Verville Sperry R-3 had three major bugs: vibration, incipient wing flutter, and poor streamlining of the retraction wells. The first two problems would continue to plague the monoplane as speeds went up. The third occurred because, although the wheels retracted into the wing, the designers were not sophisticated enough to have streamlined them into place by means of gear well doors. The wheels did retract, but the gap in the wings caused almost as much drag as if the wheels had been fixed.

A year later, the R-3 was at last fitted with the Curtiss engine and wing radiators and could attain a top speed of 233 mph, but the situation was still bad. The plane had been intended to have the Curtiss-Reed metal propeller, but another bureaucratic intervention at Curtiss's behest forced a wooden propeller to be substituted once again. It lasted only through the first lap before vibrating the R-3 out of the race.

Ironically, the R-3 came back in 1924 to win the Pulitzer against lackluster competition, after which it was relegated to a museum. The aircraft, with

Glenn Curtiss.
**BELOW RIGHT:**
There was no nonsense about rank having its privileges during the times of the Curtiss HS1-L flying boats.
**OPPOSITE:**
The beautiful brass radiator and wooden propeller of a JN-4.

**OVERLEAF:**
Some designs transcend the ages: this Myers OTW, gussied up with wheel pants and an Al Williams paint job, could represent aviation from 1918 until today.

its unrefined combination of leading edges, was doomed to make its reputation in a single race. If the Air Service had simply spent money on a development test program for the R-3, rather than on preparing for and entering contests, the United States could have been operating fighters comparable to the British Hurricane by 1932. As it was, the Verville Sperry R-3 was a lesson lost on its time.

## THE METAL MONOPLANE EMERGES

Yet the monoplane could not be denied. All over the world, engineers were examining the mounting advantages of all-metal construction and devising new structural techniques to use it. Some engineers were wedded to traditional practice—Fokker Aircraft would use the same basic methods of construction through 1939—while others valiantly sought to create all-metal aircraft. Still other engineers, such as John Northrop, were versatile enough to use both approaches successfully. Working for Lockheed, Northrop created the Vega, a

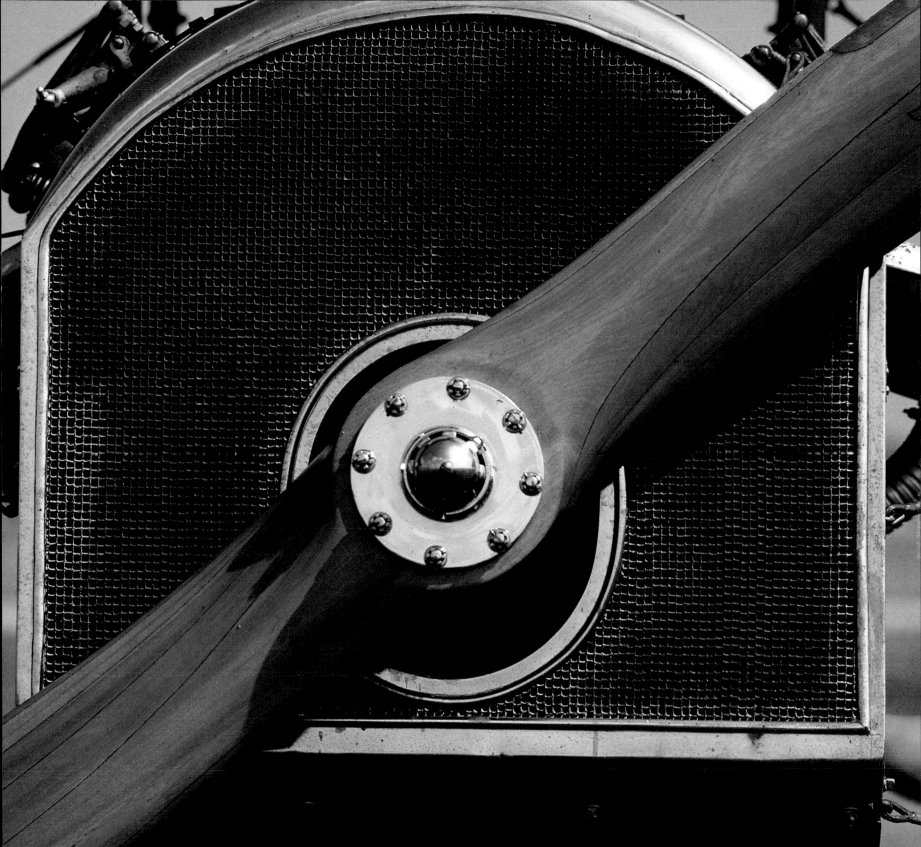

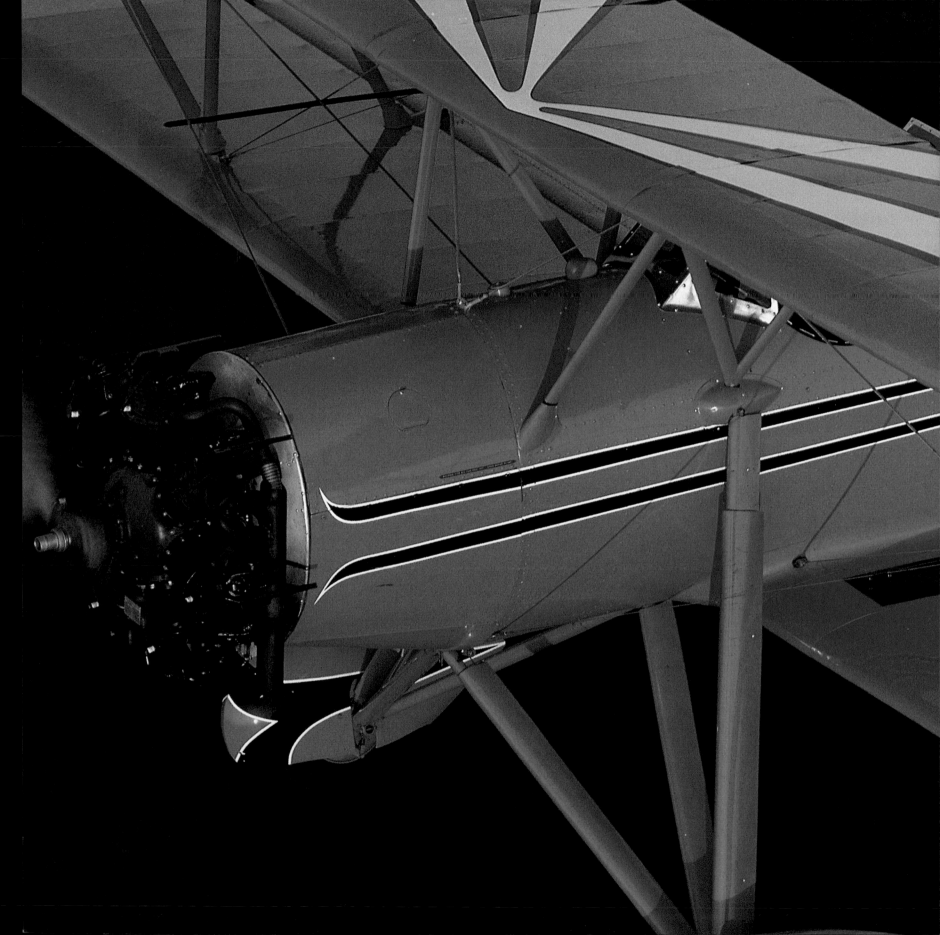

very clean all-wood monoplane that was the vehicle of many star pilots of the 1930s. The Vega series was expanded into a long Lockheed line that included the Sirius, the Altair, and the Orion; all consisted of the same basic airplane design, with the wing rearranged from top to bottom.

But Northrop saw that metal construction was the wave of the future, for reasons ranging from durability to cost of manufacture, and he created the Northrop Alpha. In the Alpha, which was to metal what the Vega had been to wood, Northrop's fine eye for streamlined beauty was evident; less obvious was his attention to strength and manufacturing costs. The Alpha featured multicellular wing construction. Each wing's interior was a series of metal angles, easily formed on ordinary shop tools and so strong that a steamroller could be driven across an assembled wing repeatedly without damaging it.

Other designers were looking to the future, too. At Boeing, the end of the company's traditional line of fabric-covered biplanes was signaled by the Monomail. Like the Alpha, the Monomail was an all-metal, low-wing aircraft that accommodated the pilot in an open cockpit, far back in the fuselage. The mail pilots preferred this arrangement, liking both the visibility it provided and the crash protection created by the long body of shock-absorbing fuselage ahead of them. The Monomail's wing construction, however, was far different from the Alpha's. It consisted of built-up spars formed of square-section dural tubing, riveted together, with Warren truss ribs supporting corrugated aluminum—not unlike the construction arrangement on the Ford Tri-motor. The corrugated aluminum on the Monomail was then covered with a smooth metal wing skin.

Only a few Alphas and only two Monomails were built, but their method of wing construction survived. The Alpha wing was specified by TWA as the type for Douglas to use on the DC-1, and the Monomail wing was used successively by the revolutionary Boeing 247D transport and by perhaps the most famous of all bombers, the Boeing B-17 Flying Fortress.

**RIGHT:**
Not radical by contemporary standards, the *Spirit of St. Louis* was transformed into a symbol of the aviation revolution (*top*). Wiley Post's Lockheed Vega had enormous development potential. Note the NACA cowling and wheel pants, developments that added as much as 20 percent to the plane's performance (*bottom*).

**OPPOSITE:**
The little Dayton Wright RB of 1920—with its retractable gear and cantilever variable camber wing—was years ahead of its time but had the usual problems: wing camber and retractable gear were actuated in unison (*top*). The many radical advances of the 1930 Boeing Monomail included a smooth skin all-metal construction and retractable landing gear (*bottom*).

## New Problems, New Solutions, New Problems

Each leading edge inevitably brought with it new problems to be solved. Retractable landing gear, for example, was often unreliable: one wheel might stick in the up position while the other extended, or perhaps neither would come down. Flaps intended to increase drag at landing could instead cause asymmetric control problems if a linkage broke and the flaps went down on one side only. In time, designers learned to make these items virtually foolproof.

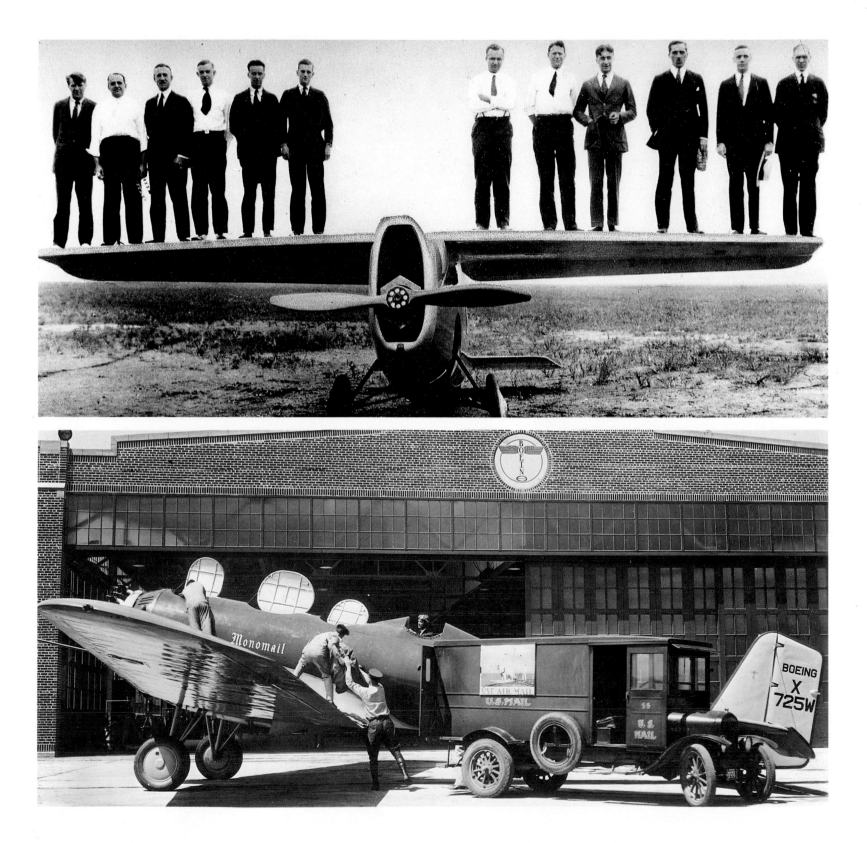

84

**LEFT:**
The airplane was noisy and accommodations were spartan, but the reliable Ford Tri-motor became a legend.
**RIGHT:**
The name Ford was synonymous with automotive quality and when applied to the Tri-motor ensured instant acceptance.

The early 1930s were thus a designer's paradise. Again and again, new designs overcame obstacles thought impassable only a few years before. The essential elements of the modern aircraft were channeled from many points of the design compass.

At this point, the radial engine had become standard in America for air transportation. Rugged, reliable, and powerful, the Wright and Pratt & Whitney engines launched American airliners into a five-decade period of supremacy; the reliable power of the engines was metered through variable-pitch propeller systems, which had just emerged and which would be as important to aircraft development as any other recently evolved device.

But big engines were not enough, and neither was the construction of all-metal monoplanes. Radial engines were perfectly flat and presented a

**FOLLOWING PAGES:**
*Pages 86–87:* A few Ford Tri-motors still fly—instantly recognizable, unforgettable.
*Pages 88–89:* During the 1930s, as aircraft shapes became more sophisticated, cockpits became more complicated. Here is a classic case of cockpit clutter.

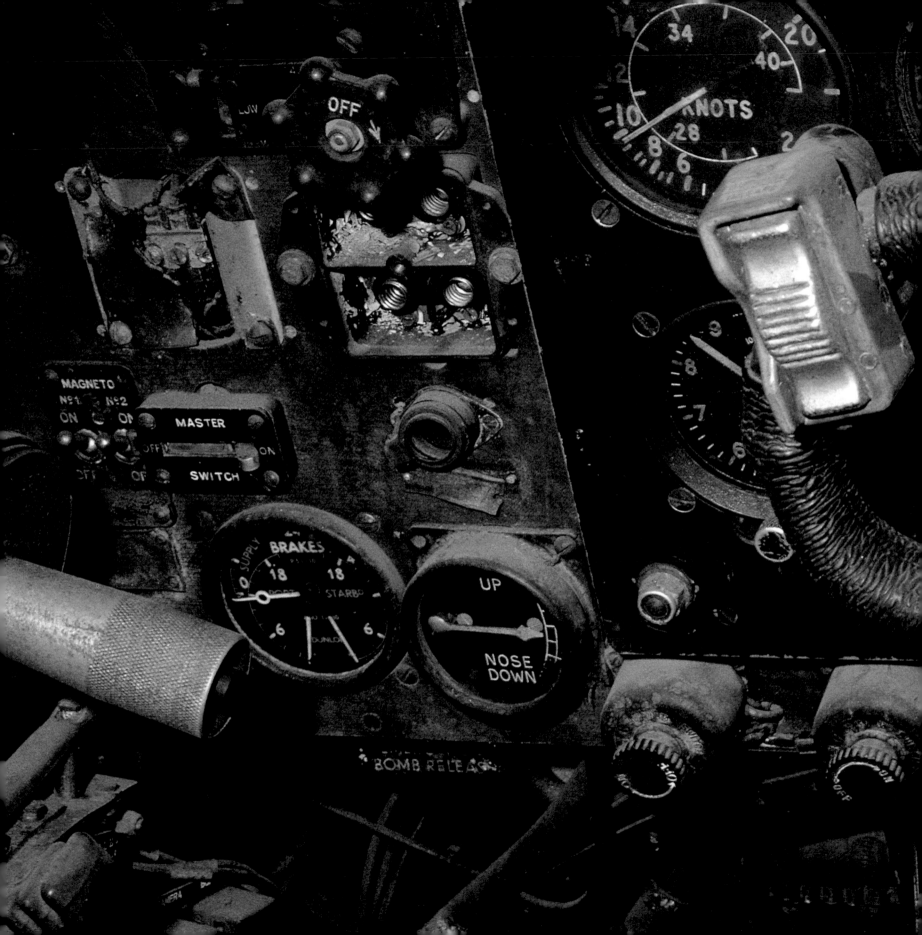

corncob series of corrugations to the wind, causing a great deal of drag. Experimentation had been done with various sorts of "speed rings," but it remained for the National Advisory Commission for Aeronautics to devise the NACA cowling.

A team led by Fred Weicks (later father of the "volksplane" Ercoupe) developed the engine cowling. Its streamlined shape surrounding the engine reduced drag and enhanced cooling, improving the performance of aircraft such as the Vega by as much as 20 percent. Similar cowlings immediately became standard equipment on high-performance aircraft.

For years afterward, engineers sought to refine this new streamlining to a point at which drag was reduced to the level attained with liquid-cooled engines. Although cooling remained a problem, such later aircraft as the lovely Focke Wulf Fw 190 did achieve drag coefficients comparable to their liquid-cooled competitors.

Curiously, while the cowling solution appeared on the surface to be simple, it was not. Not only did the internal baffling have to be structured with scientific exactness to permit engine cooling, a variety of cowl flaps had to be developed to permit regulation of the cooling. Experimentation never ceased; and no matter how blandly similar cowlings may have looked externally, on the inside they were all custom designed.

While the cowling problem looked deceptively simple, the mechanical and structural problems responsible for drag on hanging fixed gear were more immediately apparent. The first and easiest answer was to cover the wheels with streamlined fairings, ranging from small "pants" (like those used on the Curtiss Hawk P-6E) to huge "trousers" (like those of the Northrop Alpha and the Gee Bee racers). But engineers wanted more, and the ideal solution was to pick up the wheels and store them inside the airplane, as was done in the Verville R-3.

In fact, the idea of retractable landing gear was hardly new; Penaud and Gauchet had patented the idea in France in 1876. Other approaches had followed, from J. V. Martin's elegantly simple device of 1917—with its catchy slogan "Every First

RIGHT:
The name Hamilton Standard became the hallmark of propellers.

**OPPOSITE:**
All of the advancements of the 1930s came together in the handsome Lockheed *Early Bird.*

**LEFT:**
One of the earliest retractable gears to enter production was that devised for use on the Grumman XFF-1 two-seat fighter (*top*). The fully faired gear used on the 1930s Lockheed Orion folded inward into the wing (*center*). The most modern Grumman retractable gear—used on the X-29 fighter—is brought into the fuselage, this time behind smoothly faired doors (*bottom*).

Class Aeroplane Requires a Retractable Chassis" —to the advanced Dayton Wright Racer of 1920, whose gear system clearly presaged later Loening and Grumman practice. The R-3 represented yet another effort in this area. But retractable gear had never caught on because the weight and complication of the systems involved did not pay off at the airspeeds being flown.

Jack Northrop's Alpha was a case in point. Almost no difference existed between the speed of his aircraft flown as built, with fixed, faired landing gear, and an Alpha flown with retractable gear and wheels up. The difference in reliability and maintenance, on the other hand, was great— a pilot cannot forget to lower a fixed gear. Yet the Alpha was the crossover point; and in the slightly faster Lockheed Altair, the retractable gear more than paid for itself.

Manufacturers approached the problem of retractability in a wide variety of ways. Boeing invented an electrically driven system that used great worm gears to drive the wheels up and down, and the company elaborated on this later by having the wheel geared to turn 90 degrees and lie flat in the wing for easier stowage. Douglas chose a rugged, double-yoke hydraulic system; the DC-3 landing gear could be seen to struggle up in stages, the left side usually in advance of the right. In both the Boeing 247 and the DC-3, the wheel retracted to a position that still permitted it to roll—in case the pilot forgot and made a wheels-up landing! Grumman preferred a manual system that required forty-seven turns of the crank to lift or lower the landing gear, which was stowed in a paunchy belly. There were almost as many systems as there were manufacturers, for aircraft companies notoriously hated to pay license fees. (Boeing did license its gear to Curtiss, however, which built more than 15,000 P-40s using it; only a few years later, Boeing licensed the tandem gear used for its B-47 from Martin.)

### FLAPS:
### THE BEGINNING OF THE VARIABLE WING

With streamlined fuselage, more powerful engines, reliable controllable-pitch propellers, and retractable landing gear, aircraft were suddenly much faster. The Ford Tri-motor had loped along at a 90-mph cruise speed; the DC-3 was almost twice as fast. And when aircraft go faster, they generally land faster, necessitating some type of trailing edge flaps for drag and lift.

Like retractable landing gear, flaps were not new. The French Breguet of World War I, for example, had huge flaps that dropped automatically at 70 mph to provide greater lift. But flaps were unnecessary until approach speeds began to climb to 100 mph and until touchdown speeds for transports still on the drawing board were forecast to reach speeds as high as 80 or 90 miles an hour if flaps were not installed. Orville Wright's last major

The Douglas DC-3 was the first commercial air transport to make money without an airline subsidy.

**OPPOSITE:**
A wartime propaganda photo shows an eager pilot climbing into the cockpit of the super-streamlined Lockheed P-38 Lightning.

**OVERLEAF:**
Boeing revolutionized air transportation with its Model 247, the world's first cantilever wing, all-metal, twin-engine airliner.

invention (and second in importance in his career only to the Flyer itself) was the split flap, which was used on many aircraft, including the DC-3. On the DC-3, a section of the bottom of the wing could be deflected selectively by hydraulic pressure, increasing drag and lift and lowering landing speeds.

Flaps became progressively more complicated over time as they encountered the twin problems of reduced storage space and higher stress, until today a passenger in the window seat of a Boeing 727 can watch as the wing flaps unfold from a thin airfoil into almost a barrel shape, all under the smooth, powerful control of electric motors and enormous jack screws. Indeed, these intricately structured flaps make the 727 possible, since in their convolutions they provide enough lift for take-off and landing, and when fully retracted they leave only a thin, high-speed wing.

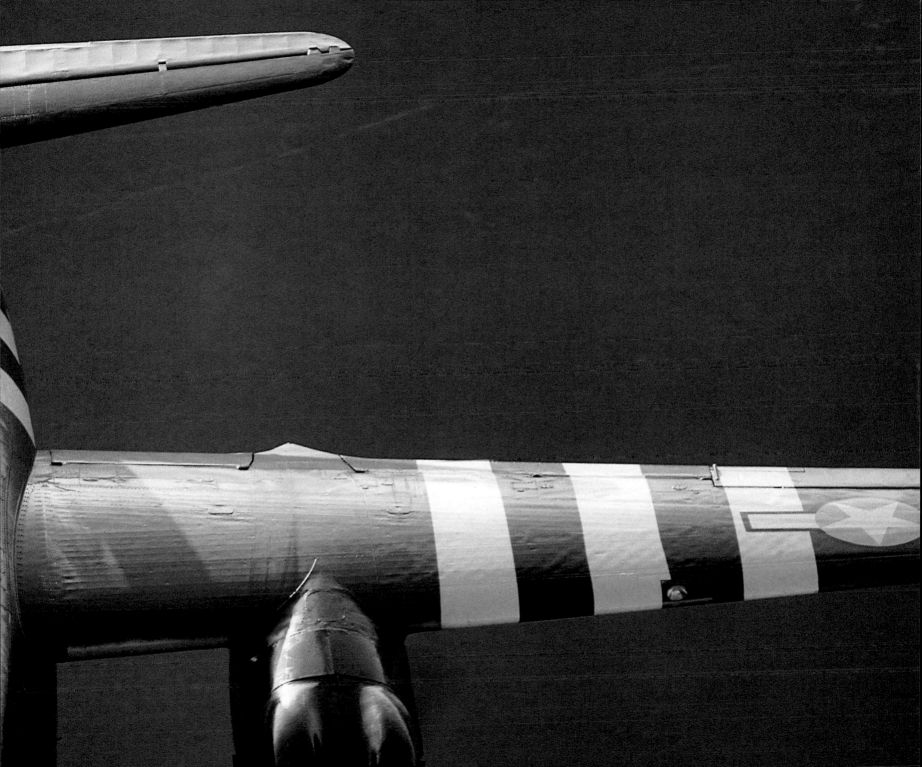

WING-SHIFT
For Setting the Wings at
the proper Trim Angle

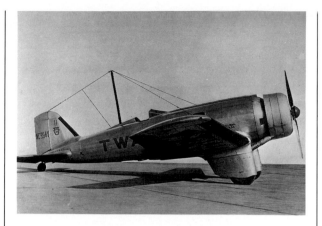

## ICING: THE GREAT LIFT DESTROYER

Accompanying these refinements was a new and major problem: icing. The hazard of ice accumulation is not weight—although sometimes this becomes enormous—but alteration of the airfoil's shape as a result of the buildup. As the airfoil becomes deformed, lift is lost and the airplane becomes progressively less airworthy. Predictably, ice affects relatively sophisticated airfoils more than it does the simpler types.

De-icing biplane transports or the corrugated skins and dangling engines of the Ford Tri-motors had not been feasible because of the excessive number of surfaces and protuberances ice could form on, and because no simple means of removing ice was practicable.

In 1932, a young man named Russ Colley made the first practical attempts at de-icing in a Douglas biplane flown by Wes Smith. Colley had fitted to the leading edges of the wing a primitive rubber sleeve which contained tubes that ran to a manifold in the cockpit. He connected a bicycle pump to fittings on the manifold and would pump up the tubes in the rubber sleeve. The flexing thus induced cracked the ice.

On this flight, Colley was considerably discomforted by ice splinters flung off the propeller and back into his face, so he also conceived the first propeller alcohol anti-icing system. Colley's ideas paved the way for the later system of de-icing

wings, which used an engine-driven pump and an automatic cycling system and formed the basis for more sophisticated systems of propeller de-icing.

The Goodyear Company—better known for its production of tires than of aviation gear, but for many years an important contributor to aviation—introduced in 1934 a de-icing boot that could be applied to the wing of a cantilever monoplane. The boot was a relatively simple device: a sleek rubber skin that was applied in tests to the leading edges of the wings (and tail surfaces) of a Northrop Alpha. The rubber skin was fitted with capstans through which air could be induced, under pressure, in a rhythmic cycle. When a coat of ice built up, the pilot would activate the de-icers; the inflating and deflating capstans would then crack the ice, and the air stream would blow it off. It was a powerful solution to what had previously seemed an unsolvable problem, and it brought regular air transportation a step closer to railroad-schedule regularity.

Propellers also ice, with even more dangerous consequences than follow from wing icing; an imbalanced propeller blade at 2,000 revolutions per minute is an immediate hazard. To offset this, prop de-icers and anti-icers were developed. A reservoir of alcohol was routed to the hub of the propellers, where it could be slung out by centrifugal force at the pilot's command. Just prior to entering icing conditions, the pilot would have a preventive coat of alcohol applied to the propellers. More often, however, a sudden uneven rumble of the propellers would be the first signal of icing. The pilot would then flip a switch and be greeted by a martinilike whiff that was fiendish for hangovers. In a moment, the ice would respond like shrapnel as it melted and flew against the sides of the fuselage.

When aircraft became larger—and especially when heat from engines became abundant—heat could be ducted through the wings for thermal de-icing. Similarly, electric heating elements were built into the propeller blades, avoiding the problem of running out of the alcohol mixture.

## Two Thoroughbreds: The Boeing 247 and the Douglas DC-3

Two aircraft that illustrate the rapid progress of technology also illustrate a number of human qualities, including brilliance, conservatism, daring, greed, and luck.

The twin-engine Boeing 247 was the first low-wing, multiengine American transport, and it revolutionized the field. It first flew in 1933, and some sixty were quickly ordered by United Air Lines. At the time, there existed a trust—what today would be called a conglomerate—consisting of Boeing, Pratt & Whitney, Hamilton Standard, and United Air Lines. The first three were makers of airframe, engine, and propeller, respectively, with the assembled package being flown by the fourth.

It occurred to the management of this group that they would be in a very advantageous position if all of the 247s were delivered to United before any other airline was allowed to buy. This would effectively give them a lock on modern air transport for two or three years. A brilliant design was thus ensnared by a greedy idea.

Trans Continental and Western Air, led by the brilliant Jack Frye, immediately saw through the gambit, and went to Douglas, asking that the Santa Monica firm build a three-engine transport with a performance equal to the 247's.

A story, possibly apocryphal, has it that a huge drawing of the Boeing 247 was placed in the Douglas design department, with a sign saying, "Like this, only better." And the Douglas engineers achieved their goal, delivering the two-engine DC-1 to TWA in 1933. Orders rapidly followed for the DC-2, and in 1936, for the DC-3. The Douglas fleet of aircraft was completely superior to the 247, which was almost immediately relegated to the less prestigious routes. Only 61 airplanes of the 247 type were produced, whereas ultimately more than 13,000 airplanes of the DC-3/C-47 type were built, and many are in service to this day.

**RIGHT:**
The lovely old Douglas DC-3—affectionately called the "Gooney Bird"—is still a major factor in aviation, with hundreds flying all over the world.

**OVERLEAF:**
Pilots loved the DC-3. It was a forgiving plane, capable of flying in bad weather, loading up with ice and still struggling on to the destination.

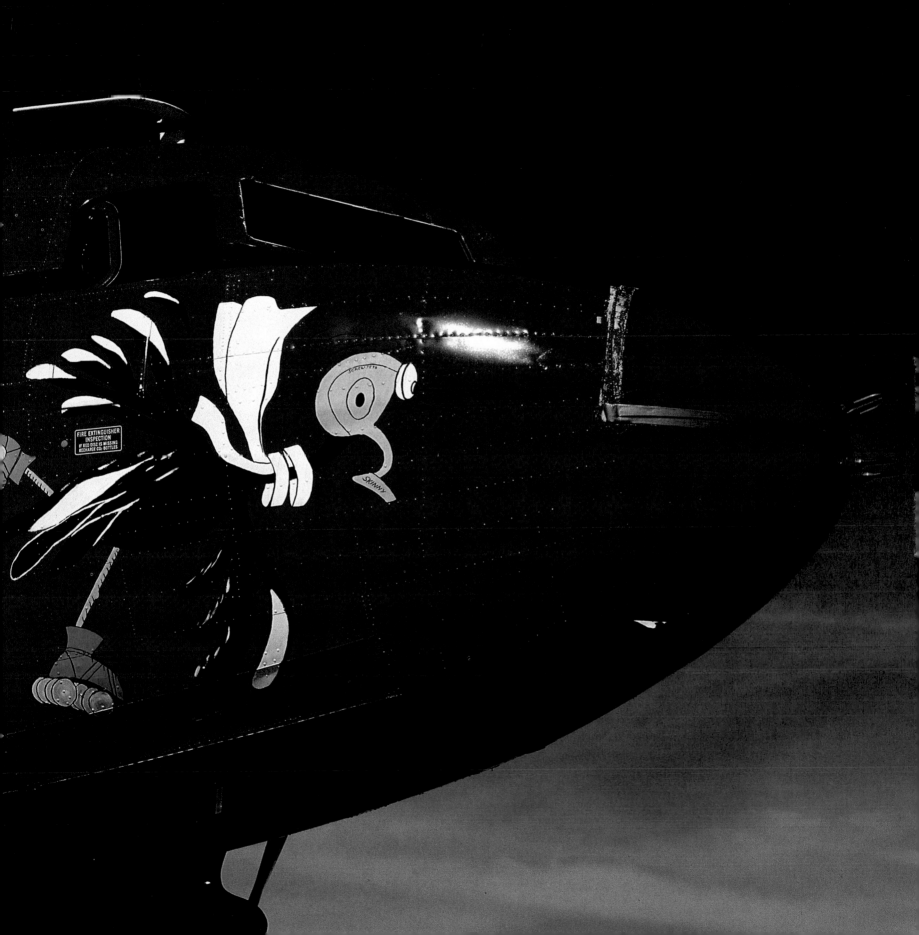

## THE FASTEST MAN ON EARTH

The year 1935 witnessed the appearance of a sleek monoplane that embraced all existing advances in one stylish, influential package: the Hughes H-1 racer.

Howard Hughes will perhaps always be remembered for his great HK-1 flying boat, erroneously (and to Hughes, infamously) referred to as the "Spruce Goose." Apart from its gargantuan size, however, the big flying boat was a technological dead end; and except for achieving maximum notoriety for minimum flying, it was a total failure.

The Hughes racer, on the other hand, was a complete success and (in aviation) his greatest memorial. Hughes was then at the peak of his mental and physical health, and he was engaged in the challenge of two difficult industries, films and aviation. In Hollywood, he had made his mark with *Hell's Angels*; in aviation, he had systematically paid his dues, from learning to fly to serving as an anonymous TWA copilot. He bought fast machines and modified them, teaching himself to fly high-powered aircraft as he set records with them.

Hughes wanted to do three things in the mid-1930s: the first was to set the speed record for land planes; the second was to establish a transcontinental record that would be beyond the reach of other aircraft for many years; the third was to win the Thompson Trophy. A survey of the market did not uncover any aircraft that could do all three things, so he characteristically decided to design, build, and fly an aircraft that would.

Again characteristically, he sought expert advice wherever he could find it, bringing in young Richard Palmer and Glen Odekirk to help him. They rented a portion of a hangar in Burbank, California, and Hughes spared no expense to build an airplane exactly as he envisioned it—an aircraft of the highest possible quality.

The Hughes racer took form as a low-wing monoplane, with bell-shaped cowling and flush riveting. The retractable landing gear was simple

**RIGHT:**
A young, untroubled Howard Hughes stands proudly behind his racer (*top*). Hughes wanted to set the world's land speed record and a transcontinental record in his racer, and he did both (*center*). He made a last great flight, around the world in record time in a Lockheed 14 (*bottom*).

and rugged, and the cockpit was enclosed. Trailing edge flaps were fitted to reduce the landing speed. The airplane, completed in 1935, was perhaps the most expensive and best-finished racer yet made. Everything was perfect. The metal skin and plywood wings were polished to a high luster; even the screw heads (which were suppressed in close alignment with the fuselage) had their slots aligned with the local airflow. The airplane could have appeared at any time during World War II and been completely modern.

Hughes wanted an extremely versatile airplane, so he designed two sets of wings for it: a 25-foot short wingspan for speed dashes, and a 32-foot long wingspan for transcontinental work.

The Hughes Flying Boat—inaccurately but inevitably called the "Spruce Goose"—was a monument to Hughes's genius, egomania, determination, and folly.

His planning paid off. On September 14, 1935, he set a new world land speed record of 352.383 mph. The long wing was then installed, and fifteen months later (on January 20, 1937), he set a transcontinental record by flying from Burbank to New York in 7 hours, 28 minutes, 25 seconds—averaging 332 mph. His third goal, the Thompson Trophy, eluded him only because he declined to enter, bowing to complaints from the racing fraternity that his competition was unfair.

The Hughes H-1 combined in one svelte package everything that aviation progress had to offer in 1935, and it did so in unbelievably beautiful style. Oddly enough, the aircraft was then flown back to California and placed in storage, with

only forty-four hours of flying time on it. Hughes kept it in air-conditioned, humidity-controlled storage until 1975, when it was transferred to the National Air and Space Museum for exhibition.

The piston-engine fighters that appeared during World War II did not differ in any substantial way from the Hughes H-1 format; and journalists rehashing the war are continually "discovering" that the Japanese Zero or the German Focke Wulf Fw 190 were copied from the Hughes. Nothing could be farther from the truth than this last claim, but it is true that the Fw 190 and the Zero were *like* the H-1 because it was so advanced. World War II

Hughes flew a Northrop Gamma 2G from Burbank to Newark in record time— 9 hours and 26 minutes— on January 13–14, 1936.
**OPPOSITE:**
One wonders if the recluse, secreted away in a dark Los Angeles hotel room, ever recalled the glittering days of sunshine and parades.

fighters were typically larger, had more powerful engines, and some were a little faster.

Over the first four decades of flight, the arts of designing engines and airframes had become interwoven, each dependent upon the other. At the end of the fourth decade, piston engines had reached their practical design limits; the next leap forward was the turbine engine revolution. Yet even this was built solidly upon the progressive developments of the past, which had seen the Wright engine's 12 horsepower multiplied by a factor of 300 as aviation progressed toward the jet age.

# 5. POWERPLANTS

Some of the most perplexed expressions ever recorded are those of laymen looking under an automobile hood at an engine that will not run.

Most of us feel anything but knowledgeable about even our familiar automobile engines, the more so since they have been incumbered with a sewers-of-Paris system of emission control devices and computer sensors. Aviation powerplants produce even stronger reactions: most people just want them to run, undistractingly and forever, modestly cowled in sleek nacelles.

Yet for others engines hold not secrets, but fascination; an entire breed of engineers deems engines the be-all and end-all of aviation, with airframes merely an appended afterthought. These engineers have been with us since the dawn of flight (Charles M. Manly and Charles Taylor, for example), and they share certain characteristics. They are precise men (and this is not sexist, for few women have created aircraft engines) who are demanding of their equipment, of others, and of themselves.

The engineers who conceive of and build aircraft engines must have a broad knowledge of physics, chemistry, thermodynamics, machining, and a dozen other skills. They must be inordinately patient, too, because engine development can take years; and when the process is along the wrong track, these are wasted years. The history

**OPPOSITE:**
Pratt & Whitney's JT8D engine powered McDonnell Douglas DC-9s and Boeing 727s.

of engine development is illuminated by brilliant products and cursed by failed programs.

The complex internal combustion piston engine, so familiar to us now in its automotive application, made aviation possible. *Internal combustion* refers to a series of contained explosions, detonated within a cylinder, that drive a piston up and down, with the piston connected to a crankshaft to deliver its power. While this is going on, myriad valves, rocker arms, cams, rollers, and other bits of ironmongery whirl about in intricate sarabands. The process is heat-intensive, making severe demands on the lubricants and the metals and requiring that every function be joined together and operate with millisecond timing.

The internal combustion engine is a machine of rocking, jolting, opposing forces; its nature is to knock, rattle, and come apart. Keeping the various forces—some reciprocal, some rotating, some centrifugal, some centripetal—all contained, while simultaneously powering them by the improbable means of igniting thousands of individual explosions per minute by means of a precise mix of gasoline and air, is extraordinarily difficult. Doing it flawlessly over hundreds and even thousands of hours under constantly varying conditions of pressure, temperature, and stress demands exceptionally fine engineering and precise machining. At stake, as we know all too well, are our very lives.

The premier early aviation engines—the Antoinnette, Renault, and Curtiss—were water- or air-cooled V-8s, relatively heavy for their horsepower but adequate for the tasks at hand. All encountered cooling problems because lubricants were still relatively simple and undifferentiated and because aircraft engines operated over longer periods and sustained relatively high output in comparison to the automobile engines from which they were descended. But the demands of flight required less weight per horsepower and better cooling; this combination was eventually put together in a reliable (for the time) package, the brilliantly executed Gnome rotary engine.

## ROTARIES

To the modern eye, a rotary engine is improbable in theory and impossible in operation. Early stationary engines required the use of heavy flywheels to dampen vibration and make the engine run smoother. In France, Laurent Seguin designed a rotary engine that at rest resembles a modern radial type; in fact, though, it was a cunning combination of transcendent vision and pragmatic common sense.

The rotary engine gave an essential boost to aviation, offering reliable horsepower at relatively low weight. It dispensed with the flywheel, and improved cooling by turning the cylinders themselves, creating a willful engine charged with the demonic energy of a chainsaw. The crankshaft was fixed, and the cylinders whirled about it; attached to a casing on the cylinders was the propeller. Typically, the engine slung a great deal of castor oil and unburned fuel from its various ports, so most rotaries were enclosed in a cowling to contain these materials. The cowlings quite accidentally helped streamline the engines and contributed to the efficiency of the airframe/engine combination.

The rotary engine had other eccentricities, as well. There was no carburetor; instead, fuel and air were brought in through the hollow crankshaft and regulated by separate valves operated by the

A three-cylinder Anzani similar to the type used by Louis Bleriot on his channel crossing. Rough-running and prone to failure, the Anzani was lightweight and thus popular.

A Bristol Boxkite replica (left) and the Eardley Billings replica (right) fly in formation. Built for *Those Magnificent Men in Their Flying Machines,* the two aircraft capture the essence of the miracle of flight.

pilot. Since there was also no throttle, the pilot regulated engine speed by "blipping" the ignition switch, shutting off spark to selected cylinders. This imparted a characteristic buzzing sound to the early rotary-powered airplanes during landing approach—a sound lovingly reproduced in the old Hollywood films on World War I even when the planes represented were not using rotaries. The blip system, combined with the fuel and oil mixture that pooled in the cowling, created an enormous fire hazard that was never far from the mind of the pilot.

The rotary engine indeed offered great power for its weight; and with good and continuous maintenance, it was very reliable. As a result, it became one of the primary powerplants of World War I, although the gyroscopic torque effect of the whirling mass of cylinders created certain control features that had to be learned and managed with studious care.

Some of the most famous Allied aircraft of World War I were powered by rotary engines, including the Le Rhone, the Clerget and Bentley, and the Gnome Monosoupape. Perhaps the best known was the "fierce little rasper," the Sopwith Camel, in which Roy Brown shot down the great Baron von Richthofen. Designed to combat the German Albatros, the Camel was a compact machine with fuel, twin machine gun armament, and pilot all closely disposed around the center of gravity. Although unbelievably maneuverable, it was difficult to master because the gyroscopic effect of the engine made turning it to the right different from turning it to the left. The Camel, like the Messerschmitt Bf 109 of a war later, was merciless with inexperienced pilots, killing them routinely in take-off accidents. Yet it became the supreme dogfighter of the war, shooting down more planes than any other.

Perhaps fortunately, the rotary engine was growth-limited; above 200 horsepower, the gyroscopic forces of the whirling metal were simply too difficult to control, so manufacturers turned to the custom-engineered adaptations of the original water-cooled automobile-type engine. On the

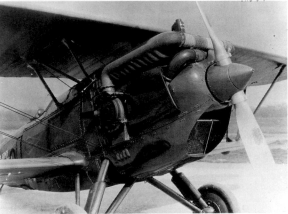

An early attempt to adjust propeller pitch.

German side, Mercedes and BMW engines became standard, while on the Allied side the French Hispano-Suiza, the British Rolls Royce, and (late in the war) the American Liberty engines were in the front rank.

## ENGINE COOLING

Liquid-cooled engines were to persist as the primary type in many fighter aircraft for the next three decades, but a number of problems were associated with them. Liquid-cooled engines (water being replaced by antifreeze/coolant fluids in the late 1920s) were heavy, and the cooling system required almost as much maintenance as the engine. The system was also vulnerable to small arms fire

**LEFT:**
The between-the-wars fighters like this 1929 Boeing P-12 used radial engines of a high power-to-weight ratio (*top*). Replicas of the Fokker triplane use either a 1930s Warner radial engine or a modern four-cylinder concealed in round cowling (*second from top*). In this replica Fokker D VIII a radial engine is installed instead of a rotary (*third from top*). The Spad XIII's engine was hard to maintain (*bottom*).

**OPPOSITE:**
The best of all worlds: WWI fighters in ultralight form. No guns, no war, just flight. From the top: the Fokker D VII, the S.E.5A, and the Spad XIII from Squadron Aviation.

—a single rifle bullet in the right spot would cause the coolant to leak out and the engine to freeze up in minutes. Nonetheless, liquid-cooled engines lent themselves to streamlining, with the cylinders fitting one behind the other in line or in V shapes that were easy to accommodate in the sleek nacelles associated with later aircraft such as the Spitfire and the Messerschmitt. In contrast, the cylinders of a radial engine are disposed about the crankshaft like spokes around an axle, and the whole presents a flat, platelike surface of considerable area to the wind.

Still, cooling an aircraft engine with water makes about as much sense as cooling a submarine's engine with air. The vast ocean of air through which the engine of a plane moved at great speeds was adequate for cooling if the engine were designed to maximize heat transfer. There were other, less obvious advantages, too. The weight of the cooling apparatus, radiator, and hoses could all be dispensed with; moreover, a radial engine was easier to work on. The U.S. Navy was very interested in air-cooled engines for shipboard operations, where liquid-cooled engines were hard to maintain, particularly at low temperatures.

## Charles Lawrance Starts the Radial Revolution

The father of the tens of thousands of radial engines that flowed from American plants, dominating the air transport fleets of the world and powering the bombers that were so instrumental in winning World War II, was Charles J. Lawrance. A modest, unassuming man (whose name is often misspelled Lawrence), he was fundamental in establishing the Wright and Pratt & Whitney engine dynasties. Once, after Lindbergh's epochal transoceanic flight behind a Wright engine, Lawrance was asked if he was upset by the lack of attention he was receiving. He replied, "No, it doesn't bother me; after all, nobody remembers the name of Paul Revere's horse."

Lawrance created the basic layout of the modern

radial engine in his J-1 engine of 1922. This leading edge of aviation engineering generated 200 horsepower, passing its reliability tests after some initial problems. The U.S. Navy was aware that Lawrance did not have the capital required to become a major supplier, and they virtually forced the giant Wright Aeronautical Corporation to acquire the smaller company. Lawrance became vice president of Wright and helped merge his advanced ideas with Wright's established manufacturing capability. This ideal business merger —of brains and money, of intuition and experience—resulted in the engine with which Lindbergh and others revolutionized the air age, the Wright Whirlwind.

The Wright J-4 Whirlwind was essentially Lawrance's basic radial engine improved upon by the engineering, mechanical skills, and long experience of Wright Aeronautical. Its successor, the J-5, generated about 220 horsepower; and after its use in *The Spirit of St. Louis,* the Whirlwind became the standard of the world and the basis for a line of Wright engines that continued until after World War II.

Frederick B. Rentschler had masterminded the growth of Wright Aeronautical, but by 1925 he found himself hamstrung by an uncaring board of directors. As happens so often, he left to found his

S. D. "Sam" Heron.
**BELOW LEFT:**
Two great men in a great airplane: Giuseppe Bellanca (left) and Charles Lawrance (right) in the Wright Bellanca aircraft.

own firm (or more precisely to resuscitate Pratt & Whitney, an old-line toolmaker in Hartford, Connecticut). He carried with him not only the basic concepts of first-class radial engines, but the most important engineers of the firm. At Pratt & Whitney, he created the first in a new line of air-cooled radials, called the Wasp, which was larger than the Whirlwind. This set off a seesaw race between Pratt & Whitney and Wright for the piston engine market that lasted until 1945.

For both companies, the path of development was not easy. Yet step by step, engine horsepower increased, and the little Lawrance J-1 that had had such a difficult time passing its fifty-hour test blossomed (through the Whirlwind and the Wasp of the 1920s) into the huge engines powering the B-29 and the B-50 of the 1940s.

## LIQUID-COOLED RIVALS

Most developments in radial engines were little known outside the industry, but they were crucial leading edges and many were adopted by manufacturers of liquid-cooled engines. The Liberty engine had been built at the height of activity in World War I, as a marriage of ideas arising at the Packard and the Hall Scott companies. Originally designed in a hotel room, the Liberty engine was manufactured in such great quantities during 1918 that at war's end they were stockpiled in Army and Navy warehouses around the country. Virtually any American aircraft other than a fighter was destined to use the water-cooled Liberty, simply because it was available.

Inherent limits in the Liberty's basic design ultimately rendered it an aeronautical dead end. But while the services waded through their stockpiles of crated Libertys, the Curtiss Aeroplane and Engine Company continued its own line of experimentation. A mustachioed engineer named Charles B. Kirkham borrowed from European practice to create a new twelve-cylinder V-12 engine, the Curtiss K-12. Kirkham's main contribution was the creation of a voluminous, hollow aluminum casting that served as a water jacket for cooling. Kirkham's

engine was later redesigned by Arthur Nutt into the Curtiss D-12, an engine that had no peer in the world. The D-12 engine powered the series of Curtiss racers and fighters, and established the firm as the most successful manufacturer in America. Perhaps even more important, however, was the manner in which the D-12 astounded the European aviation industry when it was purchased by Great Britain for use in its Fairey Fox day bombers. The D-12's preeminence was so disquieting to British national pride that Rolls Royce was given a sample of the engine and instructed to go forth and do likewise.

**ABOVE AND RIGHT:**
Charles B. Kirkham in a biplane of his own creation. Kirkham was one of the seminal figures in the development of aircraft engines, yet his name is largely unknown to the public.

**LEFT:**
Jimmy Doolittle in the Laird Super Solution, in which he won the first Bendix Trophy Race (*top*). Earlier, Doolittle had made the definitive demonstration of the possibility of instrument flights in a specially equipped Consolidated NY-2 (*bottom*).
**RIGHT:**
Universally beloved, Doolittle will always be remembered as the greatest aviator of all time (*top*). An unsurpassed race pilot, Doolittle was the only pilot both to crash a Wedell Williams (shown here) and survive the perilous Gee Bee!

## THE WAR WINNER

There was a time when the big air races—the Pulitzer, the Thompson, the Bendix, and (most prestigious of all) the Schneider Cup—received worldwide press attention and were as significant politically as the Olympics. Today, the annual unlimited class races in Reno, Nevada, are not even mentioned in the daily press.

It had been galling for the British to see the Americans emerge in 1924 and 1925 as Schneider Cup winners with the potent Curtiss airframe/engine combination and the piloting skills of men like Jimmy Doolittle and Al Williams. It was intolerable for them when the Italians advanced in 1926 to share leading honors on the strength of their bright red Macchi seaplanes and dapper pilots such as the

''three B's'': De Briganti, Bacula, and de Bernardi.

Great Britain was in an awkward position to respond, caught as it was in the depths of economic depression and governed by a Labour party notoriously opposed to expenditures for nonsocial causes. Help came in the form of a spontaneous donation of £100,000 from Lady Houston, an amount sufficient to permit Supermarine's R. J. Mitchell to design a new racer and Rolls Royce to develop a new engine. The Rolls development effort had begun with its military engine, the Buzzard, which was modified into the R (for racing) engine, which ultimately was capable of 2,550 horsepower—and had an estimated life of one hour!

The engine and airframe were combined in the Supermarine S.6 B; and this airplane not only retired the Schneider Trophy with Britain's third win

(in 1931 at 340 mph), it went on to set a new absolute world speed record of 407 mph. The lessons learned from this remarkable engine—a hothouse development that called upon all the brilliance and the finest eccentricity of its developers at Rolls—led to the Merlin engine that powered the Spitfires, Hurricanes, Mosquitos, and Lancasters of World War II.

By looking back and tracing carefully, we can see that the crucial victory of the Battle of Britain was actually made possible by mild-mannered Curtiss engineers in the 1920s. If any one of the tenuous links in the chain of circumstance that followed had been broken, Germany might well have won World War II in 1940. And the links *were* fragile: politics might have dictated the purchase of Wright rather than Curtiss engines; Kirk-ham or Arthur Nutt might have gone into business for himself and diverted his efforts from the development of the Curtiss D-12; England might have put an embargo on Fairey's purchase of the Curtiss engine, and then insisted that the competitive Napier Lion be developed; Lady Houston might have invested her money in diamonds rather than in England's honor; the happy marriage of Mitchell's airframe genius and the thundering power of the R engine might have been blocked by the fact that Mitchell was already terminally ill, with only a few more years of inspired design left to him; and finally, the Italians might have had better luck and brought their hot Macchi-Castoldi MC 72 to the race—which it would have won. So many things needed to happen and did; so many things might have happened and did not.

The development of liquid-cooled engines and radial engines followed parallel paths in increasing their horsepower over time under the impetus of competition for contracts and the predilections of company engineers. Once committed to a type of engine, companies found it difficult to switch. Pratt & Whitney was forced to dabble with liquid-cooled engines by the Air Corps, but it jettisoned them as soon as possible, in order to concentrate on air-cooled engines. Packard, despite achieving very little success with liquid-cooled engines, never strayed from the formula except for a brief foray into diesels. Curtiss-Wright abandoned liquid-cooled engines early on, while Allison never tried anything else. The same pattern developed in Europe, for similar (and forceful) reasons: engine development required too much money and time for even large firms to fritter away their efforts on different approaches undertaken simultaneously. The cobblers stuck to their lasts.

## ENGINE INTERNALS

Both liquid- and air-cooled engines depended on innovations that were often either concealed within the bowels of the engine itself or extraneous to it. One of the most important factors in both cooling types was cylinder head design: the shape of the cylinder, the placement and number of the valves, and the arrangement for cooling. (So sophisticated did the design of fin shapes on air-cooled engines become that much of the art was lost when the major companies stopped building pis-

ton engines. Yet in 1905, when Japan's Suzuki Company needed a radical new cylinder head—cooling element for its oil-cooled motorcycle, it turned to the casting and shaping techniques for fins that had been developed by Curtiss Wright more than forty years before. Sometimes an idea maintains its sharpness, needing only new frontiers to shear.)

The valves—ordinarily (to the user) one of the most anonymous of engine parts—for a time represented the weak link in engine design. Operating at the very center of the engine's repeated explosions, the valves were crucial to the proper intake of the fuel-air mixture and the almost instantaneous expulsion of the exhaust. Valve failure was common because the metals could not withstand the combination of intense heat and pounding forces. A great step forward was made when Sam D. Heron, an irascible Scotsman, invented the sodium-cooled valve, permitting reliable operation of valves at far higher temperatures than were previously maintainable.

The liquid-cooled engine received a giant boost with the development of ethylene glycol for cooling. Heron suggested that this liquid compound be mixed with water not only to serve as an antifreeze, but to permit operating temperatures well above the boiling point of water. An additional benefit of the fluid was that it reduced the combined weight of radiator and coolant by about 50 percent, with a corresponding reduction in drag. It was the perfect competition, in time and purpose, to the NACA cowling.

Both types of engines ultimately benefited from the work various companies undertook on superchargers. Dr. Sanford A. Moss of General Electric had created a successful turbo-supercharger for the Liberty engine as early as 1918, and it was used on a series of record-setting high-altitude flights. Moss's device was almost a jet engine, lacking only the latter's combustion element. In a vastly improved form, the turbo-supercharger was used on a variety of American fighters and bombers, from the B-17 to the P-38; without it, the air war over Europe would probably have been lost before it began. (Most foreign engine manufacturers preferred some variety of gear-driven supercharger.)

In a similar way, all engines gained by the improved octane rating of fuels. General Jimmy Doolittle, reknowned for his air racing skills and combat leadership, actually did more for the Allied air war effort as a scientist than he did as an airman. Almost singlehandedly, he persuaded the U.S. government and Shell Oil Company to develop 100-octane fuel and to create the refining capacity and store the necessary materials (such as tetraethyllead) to begin large-quantity production when the war began. (Later, a Shell official ruefully remarked, "He wouldn't have had such a problem convincing us, if we had known we were going to make 20 million gallons a day of it.") In contrast, the Germans and Japanese felt fortunate when they were able to obtain 87-octane fuel; the 100-octane standard used by the Allies enabled them to build more powerful engines with greater sustained combat capability.

## Spinning Edges

None of the improvements in the engines—not their increased size, improved cooling, better fuel, or greater power—would have meant anything without corresponding development of propellers. The propeller is to the engine what the paddle is to the canoeist: it converts the power into thrust.

Propellers, from the Wrights' on, had been a compromise. The angle of the propeller's spinning blade that optimized takeoff and climb was not optimum for cruise; the angle best for high-speed flight was not suitable for a go-around in case of a balked landing. During the early days of flight, many otherwise successful aircraft were dropped from use because of inadequate propellers. And the process of selecting a propeller was as empirical as the process of selecting an airfoil for the wing. If the designer did not have a fine eye for compromise, performance suffered intolerably at one end or the other of the performance spectrum.

The desired solution was seen to be some mechanical means of altering the angle of the blades in flight. The major challenge was that the whirling blade possessed enormous dynamic energy and was not easy to control. Yet engineers tried to solve the problem at an early stage in the development of flight. In World War I, each side developed proposals for variable-pitch propellers, but engineering materials and control methods could not yet solve the problem.

A controllable-pitch propeller is roughly analogous to the transmission on an automobile. In a car, we shift gears manually or automatically to match the engine speed to the speed of the vehicle. The propeller faces similar problems that are complicated by such factors as changes in air density and the sometimes supersonic speeds reached by the whirling tips of the blades.

Some obvious empirical solutions were attempted. At first, alternate propellers were produced, to be switched as necessary before flight. Then propellers were made adjustable on the ground, allowing pitch settings closer to the optimum for cruise to be selected if the takeoff field were long enough to permit them. If the field were very short, the propeller could be ground-adjusted for a fast takeoff, with a subsequent sacrifice in cruise speed.

By the early 1930s, more sophistication was obviously required; pilots had to be able to control the propeller from the cockpit, if they were to manage the existing situation. The process was gradual, with the first adjustable propellers having only two positions—one for takeoff and the other for cruise. Most often these were simple manual adjustments, made via a series of linkages.

Some methods of propeller adjustment were quite ingenious, like the one used on Michel Detroyat's Caudron C-460 racer in the 1937 Thompson Trophy race. The little blue aircraft, its 380-horsepower Renault engine a positive insult to its muscular American competitors, was fitted with a Ratier propeller pumped up by means of compressed air to a low-pitch position (a fine blade angle suitable for low-speed operation) for takeoff. After takeoff, the air automatically bled off, and the propeller blade moved into position for high-speed flight. That was it! No further adjustment was possible until it got back on the ground, but the advance was sufficient to help the Caudron win the race at 264 mph against much more powerful competing aircraft.

The Hamilton Standard Company studied the problem closely and ultimately created the best line of propellers. It began with the controllable-pitch propeller, which permitted the pilot to adjust the propeller hydraulically to any desired angle of pitch. This was followed by the constant-speed propeller, which permitted prop speeds to be set and maintained exactly, no matter what the variation was in altitude or aircraft speed. Eventually, both propellers and piston engines were brought to a height of efficiency, just in time to be superseded by turbine engines. The propeller proved to have greater longevity than its reciprocating engine partner, however—not only surviving as the turbo-prop (a propeller geared to a jet turbine engine), but reemerging as the turbo-fan so familiar today and slated to continue as the prop-fan or free-fan in tomorrow's jet engines.

## THE WHINE OF THE JET

Throughout the long development period of the powerplant and its constituent elements (fuel, cowling, and propellers), airframe design was always in advance of engine design and constantly pressed the need for more power. Many aircraft were built with the expectation that a more powerful engine as yet in its planning stages would appear at the

Sir Frank Whittle.

**FOLLOWING PAGES:**
*Pages 128–129:* The Boeing B-17 incorporated the right combination of leading edges to make it a war-winner.
*Pages 130–131:* The term "Flying Fortress" as originally applied was a terrible misnomer. Armed only with five hand-held .30 or .50 caliber guns, this plane was unprepared for war.

time of the airframe's debut; when engine development lagged, the airplane became a dead end. This was the case with such American behemoths as the Boeing XB-15 and the Douglas XB-19. Engines were getting bigger, heavier, more complex, and harder to develop. Further increases in power had become possible only through vast increases in displacement and through installation of enormously complicated systems.

By the last year of World War II, piston engines had clearly attained their developmental peak. The largest reciprocating engine to come into service was the twenty-eight-cylinder "corncob" Pratt & Whitney R-4360 (the designation refers to the type of engine—radial—and the cylinder displacement—4,360 cubic inches), which was ultimately developed to provide about 3,500 horsepower. (These rated horsepowers represent levels for continuous delivery over an extended period of time in actual field service. Racing engines such as the Rolls Royce R of Schneider Trophy fame could deliver high horsepower, but only briefly. The operational engine's task is far more demanding, given the lower degree of maintenance it receives and the service it is expected to provide under all conditions.) Other larger engines under experimentation included a thirty-six-cylinder 5,000-horsepower Lycoming, which fortunately for mechanics was never brought into service. Such forays aside, it was evident to all that the end of the piston engine road had been reached.

Not quite by coincidence but certainly fortuitously, two young men had long ago foreseen a solution to the problem. In England, Pilot Officer Frank Whittle (now Sir Frank) had a sudden vision of the future: the combination of a turbine with a compressor to create a jet engine. It was the very solution overlooked by Sanford Moss. Whittle was aware of the problematic history of gas turbines, with their low efficiencies and the lack of suitable materials for them, but he persevered and obtained in 1930 a patent on his idea.

In Germany, a little later but entirely independent of Whittle's effort, a young graduate student focused on the idea that a continuous aerothermo-

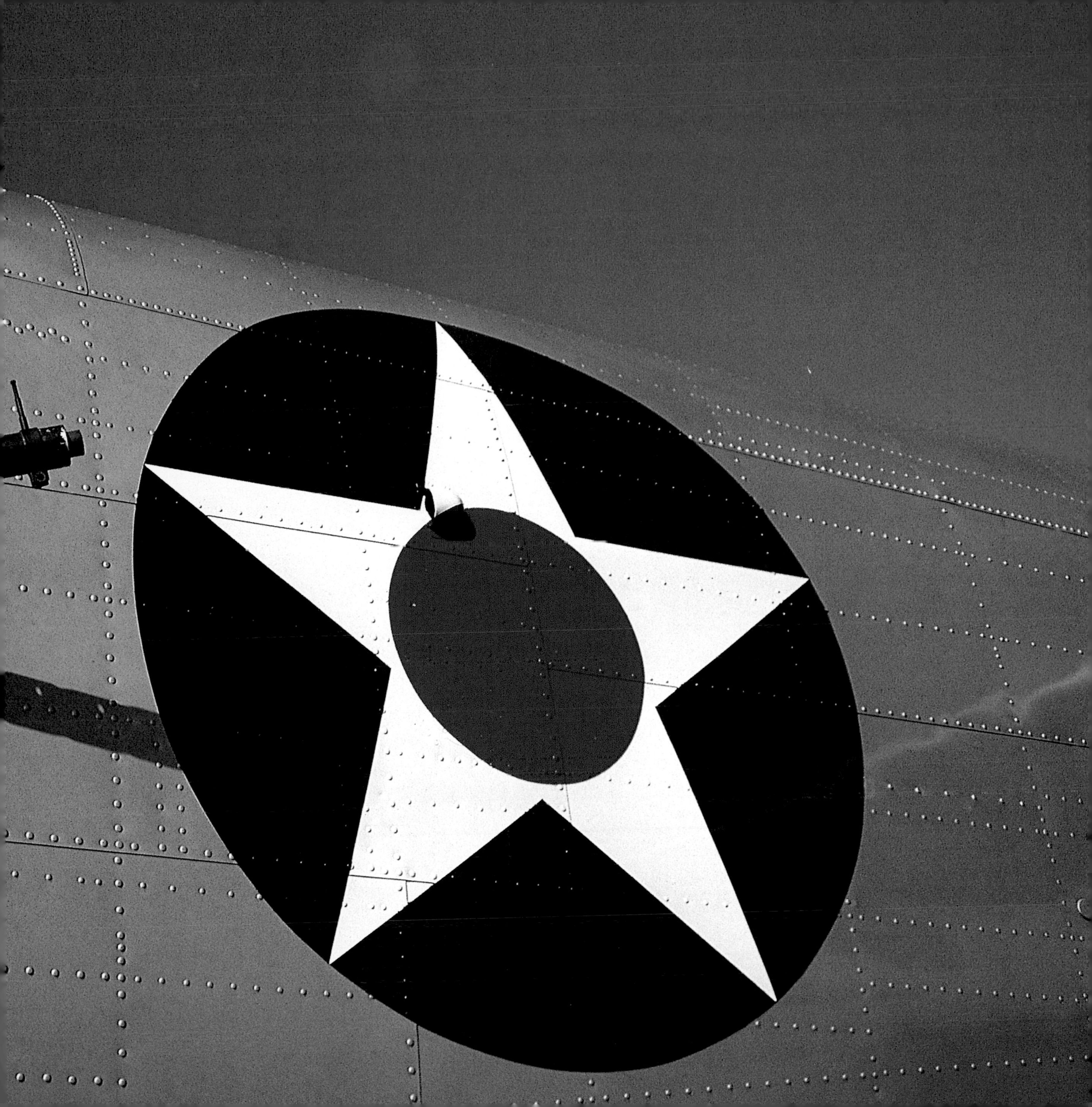

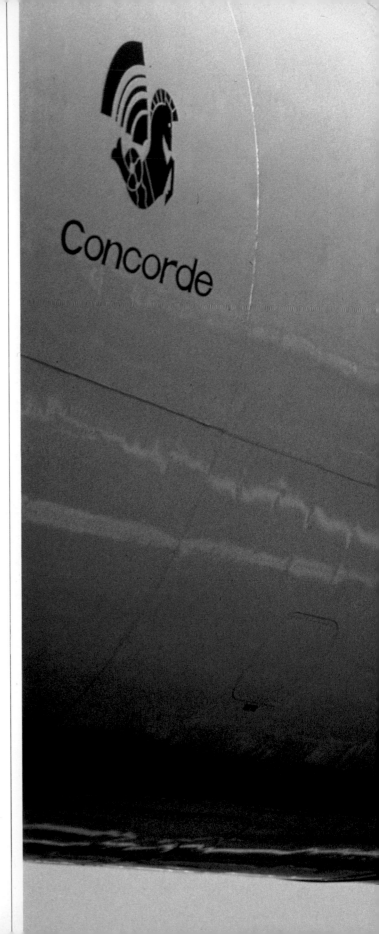

**LEFT:**
The peak of British technology: the de Havilland Comet and the de Havilland D.H. 108 (*top*).
Dr. Hans Pabst von Ohain, inventor of the jet engine in Germany. At his right is Ernst Heinkel, the controversial genius who provided von Ohain a chance to prove his theories (*bottom*).

**RIGHT:**
The Concorde was simultaneously the leading edge of technology and international cooperation.

dynamic process would be inherently smoother, lighter, and more compatible with flight than the propeller/piston-engine combination. Dr. Hans von Ohain continued to pursue the matter, and in 1934 he built a model of the engine for about $250.

The English government thought so little of Whittle's patent that it did not even put it on its ''secret'' list, permitting publication of the idea's particulars. The patent was ignored anyway, however, just as Whittle and his idea were for the next six years. Encouraged by friends to rethink his ideas, Whittle formed a private company called Power Jets Ltd. and began practical experimentation.

His research was rewarded with success in 1938 when he began to run a centrifugal-flow engine for as long as half an hour at speeds of up to 16,500

rpm. The British government at last became interested and effectively took control of the project by placing it on the secret list and tasking the Gloster company to build an aircraft (the E28/39) to fly a prototype jet engine in. The E28/39 made its first flight on May 15, 1941. Thereafter, as has happened all too often in the industry, the government failed to reward Whittle adequately, expropriating his idea and allowing others to manufacture it without paying him. A series of bitter legal battles ensued, but only served to block Whittle from receiving any financial recompense for his enormous contribution.

Von Ohain's treatment in Germany was far different, although the pecuniary result was ultimately the same. He was introduced by his Göttingen professor, Dr. Pohl, to the aircraft industrialist Ernst Heinkel. Heinkel was a great buccaneer in the

**Neither von Ohain nor Whittle nor any of their associates would have been able to believe that civilian airports would be crowded like this with a wide variety of jet air transports.**

industry—obsessed with speed and money, and capable of defying any government on any issue. He wanted to enter the lucrative field of engine manufacture but had not been allowed to do so under the Luftwaffe's ''normalization'' plans, which allocated tasks to manufacturers. Von Ohain's invention promised not only superior speed, but acceptance in an industry on the best possible basis: an exclusive product and a relatively small investment.

Heinkel provided von Ohain with direct access to the boss (Heinkel), a well-equipped shop, and unlimited funds. The result was that the HeS 3b engine was ready for flight in the specially designed Heinkel He 178 airframe on August 27, 1939. On that date, test pilot Erich Warsitz inaugurated the jet age with a simple around-the-pattern flight that was marred—also portentously—by the

The Lockheed F-104 was the first aircraft to hold speed and altitude records simultaneously (*top*). The Boeing YC-14 featured a "blown wing," jet engines mounted to force air across the wing and flap areas to improve short takeoff and landing capabilities (*center*). The Rockwell International B-1B bomber is entering USAF service after the longest gestation period of any bomber; studies began in the 1960s (*bottom*).

**OVERLEAF:**
A manufacturing technology as sophisticated as modern aerodynamics is necessary to produce modern airliners like this 747.

first jet engine failure due to a bird strike. Warsitz landed safely, however, and soon the world would be changed forever.

Exigencies of war and politics prevented the Heinkel jet engines from becoming a factor in combat. But fortunately for the Germans, a backup project—that of developing the Junkers Jumo 004 engine—had been undertaken under the leadership of a third pioneer, Dr. Anselm Franz.

Franz was unfamiliar with the idea of a jet engine and had virtually no resources, so he attempted to build the simplest possible type. He set out to build what today would be called a "breadboard" model, one on which he could learn enough to build an engine suitable for production. The pressures of a lost war caught up with him, however, and he was ordered to put his breadboard into production. Soon the engines came rolling off the assembly lines (some 6,000 had appeared by the end of the war), and they effectively revolutionized air warfare when installed in the Messerschmitt Me 262, the first operational jet fighter.

When introduced, the jet engines of all nations were cranky and unreliable. (The Jumo 004, for example, was certified for twenty-five hours and no more.) They also gulped fuel in enormous quantities; but they were light, powerful, and promised fast development.

By the time World War II was over, a generation of jet engines was under development that almost exactly matched the power of the largest piston engines in existence. Operational reliability improved rapidly, and within a few years the jet engine had completely reversed the old equation. Engines now led airframes in development, generating enough power to fly an aircraft to destruction in level flight.

The next forty years would be marked by the continuing efforts of airframes to take advantage of power capabilities made possible by repeated advances in engine design. The results of these new leading edges have been speed and more speed, changing the social, technical, economic, and political appearance of the world beyond all imagination.

# 6. THE FURTHER
## DEMANDS FOR SPEED

Some leading edges emerge almost by magic; poured from many crucibles, honed on many stones, a final product appears that is far different from and far better than anything envisioned by the original designer. Such is the case with the North American P-51 Mustang, certainly one of the most famous and perhaps the best piston engine fighter of World War II. The Mustang continues to be operated by many sports pilots, and it appears regularly in unlimited-class air races.

This outstanding fighter was conceived almost accidentally, had several fathers, and reached its standard of excellence by a fortuitous and harmonious melding of disparate developments.

North American Aviation had never built a world-class fighter and had no design for one when it was approached in late 1939 by the British Purchasing Commission to subcontract production of Curtiss P-40s for the English war effort. North American demurred, saying that it was not interested in producing a design that was already out of date, and claiming that in a relatively short period it could produce a new fighter with superior capabilities.

By April 1940, North American had arranged a deal with the Curtiss Wright Corporation to purchase wind-tunnel data and the details of a new radiator arrangement that had been used on the

**RIGHT:**
The smoke trails express the formidable drag of the old biplanes.
**OPPOSITE:**
The North American P-51, a compendium of leading edges at its debut, is still active as a racer and acrobatic plane.

Every detail of the Mustang was finished perfectly, from the almost jaunty bubble canopy and the sleek laminar flow wings to the functional beauty of the retractable gear—even to the lethal openings for the six .50 caliber machine guns in the wings.

Curtiss XP-46 (a much-improved version of the P-40 that was nonetheless vetoed by the U.S. Army Air Corps because it would have disrupted production lines).

North American's designers took this material and laid out an aircraft of substantially the same proportions as the XP-46, but incorporating the recently developed NACA No. 66 airfoil section for the wing. This section—termed *laminar flow* because it held layers of air as though laminated together in a perfectly smooth flow—had exceptionally low drag characteristics that were destined to contribute to the airplane's long flying range.

The North American team, working sixteen hours a day, turned out the first Mustang in just 102 days; it was an obvious and immediate success. The British bought 400, 2 of which went to the United States Army Air Force under the designation XP-51. Ultimately, almost 15,000 were built, and they were responsible for destroying 4,950 Axis aircraft in the air and countless more on the ground. Almost every Western air force eventually used the Mustang in one capacity or another.

The Mustang was originally powered by an American Allison in-line engine; the Allison was satisfactory for low-altitude work, but it did not develop the horsepower required at higher altitudes. The next step in the multiple-crucible process came with the installation of the superb British Rolls Royce Merlin engine into the American-designed airframe. The resulting combination, ordered into service as the P-51B, immediately established itself as the premier USAAF fighter and presented the German Luftwaffe with an unsolvable problem.

The American daylight bomber offensive had been in desperate need of a long-range escort fighter, and the Mustang filled the bill. It was the only piston-engine fighter of the war that had both the range to escort bombers all the way to Berlin and the speed and maneuverability to out-dogfight the German opposition. Hermann Goering reportedly said that he knew the war was lost when he saw the American escort fighters over Berlin.

Yet this farthest advance of piston-engine fighters was already obsolete, overtaken by the extraordi-

nary application of the Junkers Jumo 004 jet engines to the Messerschmitt Me 262. The Me 262 was the most advanced fighter of the war—a singing sword cutting to the future in a strange mélange of brilliant intuition, colossal mistakes, sabotage, and genius.

The Messerschmitt Me 262 had been conceived early on as a mere backup. The war was supposed to be won with Messerschmitt Bf 109s, but there were resources enough to experiment. No real need existed for improved combat craft when the Me 262 was first offered to the Luftwaffe in the palmy (for the Nazis) summer of 1940, and it was ordered more as a curiosity than anything else.

The original design called for a straight-wing aircraft, with twin BMW engines set in the middle of each wing. Over time, the engines used grew so large that they had to be suspended in pods below the wing, and thereafter they grew so heavy that the wing had to be angled back to provide a correct center of gravity. Thus the breakthrough design of swept wings—the most distinctive element responsible for placing the Me 262 in the technological forefront—was an engineering accident.

By the time the aircraft first flew, in the summer of 1943, the war had drastically worsened from Germany's point of view. Generalleutnat Adolf Galland, a famous ace and a grand gentleman,

**LEFT:**
In 1947, the most influential jet aircraft of all time appeared: the Boeing XB-47 (shown here in a rocket-assisted takeoff). It was the direct inspiration of the world-wide fleet of jet-powered swept-wing military and commercial aircraft of the next four decades.

seized upon the fighter as his country's only possible salvation; and it might have provided that salvation if confusion and ill-conceived priorities had not kept it out of mass production until mid-1944. By war's end, more than 1,300 had been built, but only about 300 ever saw combat.

The Me 262 thus did more for the morale of Germany's engineers and pilots than it did for the war effort. At a time when the end was evidently near and when the horrors of the Nazi excesses were being uncovered, the Me 262 leaped forward a generation in design, influencing aircraft around the world.

## THE INVISIBLE BARRIER

During World War II, fighter pilots diving at high speeds would sometimes find their controls locked in a viselike grip or would be battered mercilessly as turbulence flailed the stick back and forth in the cockpit. They were encountering the previously unknown phenomenon of compressibility drag. At higher airspeeds—given the airfoils of the day—the drag coefficient would go up sharply, and a boundary layer separation would occur, creating a wide, turbulent wake. In lay terms, this was "the sound barrier," because it occurred at the speed of sound (760 mph at sea level).

When the war ended, teams of Americans descended upon Germany to pick up engineers, aircraft, drawings, and anything else that might be useful. One of the visiting teams included R. T. Jones and Theodore von Karman. Jones earlier had written a paper predicting that sweeping the wings would delay compressibility drag, but the paper had been rejected. Now he found his theory corroborated by German practical experience. Wires were sent to the United States, and both Boeing and North American were instructed to incorporate wings of 35-degree sweep in their new aircraft, the XP-86 and the XB-47. The XP-86 (the North American Sabre) became for the Korean conflict what the Mustang had been for World War II—the premier fighter, much loved by its pilots. The XB-47, perhaps the biggest and least expensive gamble ever taken by the Air Force—the program was budgeted at only $10 million—became the most important jet aircraft of all time.

The Boeing Company had undertaken the usual process of creating an extended series of designs for a jet bomber to succeed the fleets of piston-engine B-29s and B-50s, and it had done so without startling results. Early versions simply featured jet engines hung on the B-29, yielding the worst of both worlds: a piston-engine airframe and the fuel-guzzling of the jet. But by adopting the German concepts of swept wings and podded engines, Boeing engineers developed a unique new aircraft that granted the United States strategic air supremacy for more than a decade.

While the XB-47 (X for experimental; B for bomber; and 47 for being the forty-seventh bomber type developed for the Air Force) was being built in secret in a guarded factory, it was referred to as a sacred plane: the first remark a visitor usually made upon seeing it was "Holy Christ." It was completely different from anything ever seen before—a swept-wing stiletto, with a total of six podded jet engines hanging below the wings and a funny-looking bicycle-style undercarriage. It was capable of traveling at 600 mph at high altitude and was virtually immune from interception. To extend the B-47's range, Boeing also developed a new system of air refueling, a system that used a patented "flying boom" to transfer fuel from tankers to bombers. (Air refueling had been tried as early as the 1920s, but not until the flying boom—with its ease of control and high rate of fuel transfer—became available did the tactic attain a functional level of dependability.) With air refueling, the B-47 could range the world.

The Air Force ordered a series of production models (labeled B-47A through E) over the years, and ultimately more than 2,000 B-47s of various models were brought into the Strategic Air Command (SAC) and placed in the hands of highly skilled, well-motivated crews. The Soviet Union had no counterpart bomber, nor did it have an interceptor force adequate to combat the SAC fleet. The B-47 fleet thus established a Pax Ameri-

Sir Geoffrey de Havilland.

cana that lasted until the advent of the intercontinental missile.

Beyond its military importance (leading as it did to the B-52, the most important jet bomber of the last thirty years), the B-47 paved the way for swept-wing commercial air transport. From it descended not only the line of KC-135 tankers that are still in constant use throughout the world, but also the 707, 727, 737, 747, 757, and 767 series of jet airliners with which the United States (and Boeing) established a commercial air dominance that is only now being challenged.

These leading edges—Mustang, Messerschmitt Me 262, Sabre, and B-47—all had a great deal in common: each, at the time of its debut, was the highest expression of the state of the art; each was an excellent aircraft to fly; each drew upon the best the past had offered, and then conferred something entirely new and previously nonexistent; each was beautiful, a work of art; and most important, each laid the foundation for the next leap forward.

## BREAKING THE BARRIER

Aviation was poised at a threshold: it now had powerplants that could thrust airframes past the speed of sound, but it had yet to devise airframe shapes that could break the sound barrier safely. Thus an important leap still had to be made: the actual breaking of the sound barrier. Writers in popular magazines periodically depicted the sound barrier as an impenetrable wall against which aircraft would crash and destroy themselves. Most engineers knew this was not so and regarded the problem merely as one compounded of power and airfoils. Yet the element of danger was there, as was proved when young Geoffrey de Havilland, son of the famous English aircraft manufacturer, died while attempting a speed record in the de Havilland D.H. 108 Swallow on September 27, 1946.

Robert Wood, a veteran aircraft designer, committed the Bell Aircraft Company to making a research aircraft (the Bell X-1) that would be capable of breaking the sound barrier. It was well known

**LEFT:**
The McDonnell Douglas
F-15A Eagle combined the
most sophisticated tech-
nology with the lessons
learned in Vietnam.

**RIGHT:**
In the thirty-seven years
that passed between the
first flight of the B-17
prototype in 1935 to the
first flight of the Eagle in
1972, technology changed
drastically.

**OVERLEAF:**
The helmeted crew mem-
bers of a Boeing B-52.

that .50-caliber bullets exceeded the speed of sound—760 mph at sea level—so the fuselage of the X-1 was given the shape of a bullet. For simplicity of construction and improved flight control, the design team elected to use very thin straight wings. The power was to be supplied by a rocket engine, primitive by space shuttle standards but ferociously powerful (and unpredictable) for the time.

The whole program was fraught with hazard. Streamlining, designed to avoid the shock waves associated with supersonic flight, had reduced the pilot's canopy to a glassed-in portion of the fuselage, and visibility was very poor. No ejection seat, no heating, and no other amenities of ordinary flight were provided for in the X-1. The airframe was in fact little more than a flying fuel cell, to be dropped at altitude from the bomb bay of a carrier plane. As had often happened before, the leading edge of flight demanded a great deal of courage from its practitioners.

The Air Force conducted a very systematic program—first with the contractor, and then with its own pilots—and on October 27, 1947, Captain Chuck Yeager brought the Bell X-1 (named Glamourous Glennis, in honor of his wife) through the sound barrier safely and almost without event.

The ensuing forty years have been a riot of invention, a paradise of hardware, all solidly founded on the leading edges expressed in the features of the B-47 (with its swept wings and jet engines arranged in a supremely useful configuration) and the Bell X-1 (with its supersonic design features); variations on these two themes were developed all over the world. The North American Super Sabre, the F-100, was the first of an array of supersonic American fighters that has culminated in the current F-14, F-15, F-16, and F-18 series. The Soviet Union has responded with a classic line of MiGs, from the 19 to the 31. France has marshaled its Gallic grace into a superb line of Dassault fighters, while England fielded first the Lightning, and then (in cooperation with its European allies) the Jaguar and various Tornados. Even smaller countries, such as Sweden (with the Viggen) and Israel (with the Kfir), have kept pace.

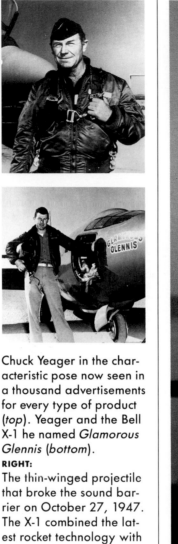

Chuck Yeager in the characteristic pose now seen in a thousand advertisements for every type of product (*top*). Yeager and the Bell X-1 he named *Glamorous Glennis* (*bottom*).
**RIGHT:**
The thin-winged projectile that broke the sound barrier on October 27, 1947. The X-1 combined the latest rocket technology with an almost intuitive airframe design: the fuselage was shaped like a bullet because bullets were known to be supersonic.

**LEFT:**
The de Havilland Comet was a sleekly beautiful aircraft that on May 2, 1952, introduced 500 mph jet performance to airline service on the London-Johannesburg route.

## THE MODERN AIRLINER EMERGES

Civil air transports developed at a slightly lower rate. The demand for performance in civil aviation has an economic foundation, and private passenger aircraft are flown for longer periods of time than are military aircraft. The beginning of civil jet air transportation was marked by tragedy, however, when the de Havilland Company tried (with near success) to steal a march on the world with its beautiful de Havilland Comet airliner—the first jet airliner to enter commercial service.

Since 1935 the world's airlines had become captive to the strides made in American piston-engine aircraft. This process had started with the DC-3, had been reinforced by the DC-4, and had been perpetuated by the DC-6 and the Lockheed Constellation. De Havilland was determined to break this iron grip, and began the process on March 2, 1952, when scheduled flights with the Comet began on BOAC's London-to-Johannesburg route.

The Comet was an instant success. With a passenger capacity of only thirty-six, it was small by today's standards but perfectly suited to the traffic. Its most important characteristic was that it could fly at a speed of 500 mph at an altitude of 40,000 feet, above much of the bad weather; this enabled the Comet to meet its schedules with admirable regularity. Encouraged as various airlines (including the giant American carrier, Pan Am) began queuing up to buy, de Havilland designed larger variations. Eventually ten airlines employed the Comet.

Then tragedy struck. A Comet broke up after taking off from Rome on January 10, 1954, crashing in deep water off Elba. An investigation was conducted, and all Comets were grounded for inspection, but no structural problems were identified, and investigators accepted sabotage as the most likely cause of the crash. Then on April 8, another Comet disintegrated—this time near Sicily. Ironically, the second tragedy occurred on the very day that salvage efforts were abandoned at the Elba crash site. Work at the Elba site was immediately resumed, and enough evidence was found to pinpoint the cause of the disaster.

The seeds of the Comet tragedy had been sown in the early days of flight. The need to pressurize aircraft so that pilots and passengers could breathe comfortably was recognized as soon as aircraft began to carry people at altitudes above 10,000 feet, where the atmosphere is noticeably thinner. The first experiment on record was at McCook Field in the early 1920s, when Captain Harold R. Harris made a flight in the pressurized cockpit of a modified Air Service aircraft. The underlying idea was simple: the passenger compartment was constructed to be strong and airtight, and air was then pumped in from outside, causing air density during flight to remain at a comfortable level.

Experimentation continued, but the first true pressurized passenger aircraft was not designed until 1935, when Lockheed created the XC-35 and won the Collier Trophy with it. Boeing built the first pressurized airliner, the Stratoliner, in 1939; and during the war, many types of aircraft were equipped with pressurized cabins—including the B-29 and the Constellation.

When de Havilland designed the Comet, its planners followed all of the accepted practices of the time. Unfortunately, these practices did not properly take into account the effects of metal fatigue on cabins being run through their pressurization cycle several times per day. The state of the art was simply inadequate.

Designers had never encountered a situation in which the entire shell of the fuselage was subjected five or six times per day to a change in pressure from sea level to 35,000 feet and back again. Their engineering data did not tell them that certain types of metals used routinely in aircraft construction were more susceptible to the process of fatigue than others. Neither did their data tell them that metal that had been formed into certain shapes was more vulnerable to fatigue than the same metal formed into other shapes. High-tensile-strength aluminum (valued because of its light weight), for example, had a tendency to fracture; and square openings—windows, hatches—were

more apt to fatigue than round ones. The fatiguing process was identical in kind to breaking a wire coat hanger by bending it back and forth; the only difference was that each bend of the cabin walls of the Comet, under the influence of pressurization and altitude changes, occurred imperceptibly. But the cumulative effects of fatigue were the same, and ultimate failure of the metal was inevitable.

The Comet was the leading edge of air transportation in every way but one: that of cabin design. The two disasters, however, resulted in all Comets' being pulled from passenger service. The firm very bravely fielded an improved model, the Comet IV, and BOAC very bravely bought nineteen. They served well, but de Havilland's lead over the American competition had been lost (apparently forever) to a sleek transport emerging from the Boeing factory—the prototype 707.

## THE IMMORTAL JET LINER

The first 707 (technically, Boeing Model No. 367-80) made its maiden flight on July 15, 1954, just three months after the second Comet disintegrated off Sicily. A little more than three years later, on October 26, 1958, Pan American Airways inaugurated transatlantic service with the 707. The jet age had at last overcome its false start with the Comet.

The 707 and its rival, the Douglas DC-8, wrought enormous changes in the social fabric of the world. Passenger ships followed the passenger train into oblivion. A transatlantic flight, once the hallowed preserve of wealthy businessmen and ambassadors, became routine. The luxurious service of the Flying Clippers of the past—with its well-cooked meals, wine service, and linens—was gone; in its place hurried harried flight attendants, dropping off plastic trays filled with plastic food. Oddly enough, the passengers did not mind, or rather they recognized that the enormous speed at which they hurtled across the Atlantic (in less than eight hours) was ample compensation for the lack of personal attention.

It is a tribute to the original Boeing design that

the very first 707s reached a speed not vastly different from that of the very latest models. The first 707s cruised at a speed of 550 mph; today a 747 is capable of 625 mph, but usually flies at about 550. The reason, of course, is that the jet engines on the first 707s already permitted them to cruise in the high subsonic speed range. To go faster would be to push at the boundaries of supersonic flight, where fuel consumption increases markedly. And such are the effects of departure and arrival procedures that an increase in cruise speed of 30 or 40 mph (say, from 555 to 585) would have little impact on the "block time"—the total time from takeoff to landing.

After the 707, industry-wide leading edges in air transportation are less evident. For example, once the 707 had been introduced on all international and transcontinental lines, the need for a short-range jet in the marketplace was clear but at the same time nonuniform. The market defined by some airlines demanded a four-engine aircraft capable of taking off from high-altitude fields like Denver, while for others it required a twin-engine aircraft capable of taking off fully loaded from the very short La Guardia runway 4-22 in New York. Whatever the specific needs of the carrier, the aircraft had to handle a passenger capacity of up to 129 tourist-type seats.

Boeing, far from being wealthy from its long run of 707 sales, was in perilous financial condition. In order to continue selling the 707, Boeing customized planes bought in quantity to suit the special needs of customers; continual modifications of this sort, however, raised the break-even point on production higher and higher. To stay competitive, Boeing had to have a new model. The engineers and planners, with their customary cool gambler's skill, stood at the design table and rolled out the 727.

This was a radical aircraft, with three engines (the first in America since the Ford Tri-motor), a tall T tail, and extremely high wing-loading. To meet the demands of high-altitude fields and short runways, Boeing devised the period's most efficient lift system: a triple-slotted flap and leading edge (in this case, *leading edge* refers to the foremost part of the wing) slat combination. At rest, the wing has a conventional appearance—swept back, but otherwise not extraordinary—but when the lift devices are deployed, the wing seems to come apart as leading edge slats move out and flaps articulate back until the entire length of the wing describes a virtual semicircle.

The result was a runaway success; the high lift at low speeds and the low drag at high speeds of the 707's articulated wing, together with the efficiency of its aft-positioned engines, provided such outstanding service that it became the most-produced jet transport in history. In all, 1,831 were built.

Competitors to the Boeing products did develop. The Douglas Company opted to build a slightly smaller aircraft, the DC-9, which was less radical and almost as successful. Boeing countered with its own 737, and the two companies continue to battle for similar markets.

Airlines historically have made projections of passenger traffic in determining the type of aircraft they need. In the mid-1960s, airlines decided that the aircraft of the 1970s would require a passenger capacity of from 350 to 375 to suit market needs. A basis for aircraft of this size could be found in the U.S. Air Force competition for the C-5A. Lockheed won the competition for the C-5, but Boeing gained the necessary data to have confidence to build the world's first high-speed, wide-body passenger jet, the huge 747.

With the 747 and its smaller wide-body cousins (the L-1011, the DC-10, and the Airbus) came a further phase in the revolution in air travel. New York-to-Tokyo had changed from a month-long sea voyage to a five-day endurance test in piston-engine airplanes to a two-day contest in ordinary jets to an overnight polar hop in the 747.

But distances and the eternal quest for speed demanded more. Transport performance had followed military performance since the beginning of flight; if fighters were going supersonic, bombers could not be far behind—and if bombers, why not then passenger aircraft?

It was an enticing line of reasoning, and soon volumes of philosophy had been published proving the inevitability and the wisdom of supersonic passenger traffic. Great graphs were constructed that showed an absolute parallel between fighter speeds, bomber speeds, and transport needs—the only differences being in the time intervals separating their introduction.

## SUPERSONIC TRANSPORT AND BEYOND

The revolution in air traffic made another great point. It was all very well for vacationers to loaf along at 600 mph, taking up to eight hours to get to New York from London, but what about the busy businessperson, for whom time was money? Was it not evident that a three-hour trip (in which a traveler would arrive in New York fifteen minutes earlier—by the clock—than takeoff from London) offered invaluable saving of business time and talent?

Still another argument was the "twelve-hour cir-

cle" that could be drawn around any departure point. Studies showed that virtually all traffic occurred between points within twelve hours' flying time of each other; beyond the perimeter of a given twelve-hour circle, little demand existed. The conclusion was obvious: supersonic aircraft would place the whole world within a single twelve-hour circle, and traffic would undoubtedly climb.

But the real reasons for the air transport industry's keen interest in creating a supersonic transport were national pride and the simple fact that it could be done. Airframe and engine technology had reached a level at which a development program might be able to produce a successful supersonic aircraft. The Russian experience with the ill-fated TU-144 had already demonstrated that, while an SST could be created, it would not necessarily be a success. The TU-144 had been the first to fly (on December 31, 1958), the first to crash (in 1973), and the first to be dropped from use (in 1978).

Once again, as with the Comet, the English were first off the mark, in early and eager collaboration with the French. The design studies for the BAC Type 223 in England and for the Sud Super-Caravelle in France had yielded almost identical configurations, and in 1962 an agreement was signed between the two nations to work together formally. The agreement was unusual in that no provision was made for cancellation; both nations committed themselves to completing the task, and the result was the aptly named Concorde—a gracious concession by the English to the French spelling.

That more people had to be involved to create a leading edge in the supersonic age than in the early days of flight was one of the many lessons of the Concorde program. Two governments, two design, assembly, and test centers, more than 800 subcontractors and suppliers, and 24,000 workers at levels ranging from engineer to custodian were necessary to give birth to the Concorde—and English and French companies employ far fewer engineers and workers than do their Russian and American counterparts. The leading edge truly has passed from the conceptual genius of an

rials. Each country had its own long-established standards for everything from sheet aluminum to fasteners to rubber grommets, and real diplomacy was needed to find common ground.

The Concorde was and is a highly successful aircraft, especially when measured by passenger satisfaction. Champagne rolls down the throats of Concorde passengers almost as fast as fuel rolls through its engines. But in this last circumstance lay the supersonic transport's undoing.

At the same time that brilliant engineers in France and England were creating the fastest and one of the most beautiful transports of all time, representatives of the oil-producing nations were hatching their new cartel. When oil prices zoomed, the Concorde lost any chance at profitability, and with it any chance of sales to carriers besides the flag airlines of France and England. It flies still, and in certain instances it has been able to pay for its operating cost; but recovery of the research and development and production expenses is not possible, and even the prospect of such gains has been dropped from consideration by all but those politicians who opposed the project in the first place.

Enormous interest had arisen within the United States for the supersonic transport, coinciding (unfortunately for it) with burgeoning concern for the environment. The supersonic transport project was too large for any individual company to handle, and as a result NASA worked in close cooperation with both Boeing and Lockheed to develop a portfolio of designs from which a production type could be chosen. The FAA was also brought into the arena, to ensure that the SST under development met commercial aircraft requirements.

Both Lockheed and Boeing investigated a whole series of configurations, with Lockheed tending to prefer an ogival delta shape and Boeing favoring the swing wing. Much data and flight experience were available on both types—from the B-58 and XB-70 programs for the delta shape, and from the F-111 program for the swing wing. Boeing won the competition with its swing wing design, but in the subsequent development process it finally elected to proceed with a modified delta shape.

individual to the collaborative genius of a scientific bureaucracy.

The British and French agreed on a formula for the Concorde: a cruising speed of Mach 2.0 (about 1,250 mph, or twice the speed of sound); seating for approximately 130 passengers in a long, slender tube of a cabin, arranged in sets of two seats on each side of an aisle; a nose that "drooped" to provide better pilot visibility at takeoff and landing; four Rolls Royce/SNECMA Olympus engines; and a structure made primarily of aluminum alloy. A combination of economy and the thermal limits of aluminum dictated the Mach 2.0 speed.

The cooperation between the two countries was enormously successful, despite differences in language and measurement systems. A more difficult task involved developing joint standards for mate-

American ambitions for the SST were far greater than British-French ambitions for the Concorde had been, but some limits had to be observed. Design studies indicated that Mach 2.0 was the speed limit for an aluminum structure, and that Mach 2.7 was the speed limit for an aircraft that did not have a means of cooling its fuel.

The temperatures generated by flying at supersonic speeds are enormous; the Iconel X nickel of the hypersonic experimental North American X-15 would turn cherry red from air friction at the craft's maximum speed. The X-15 flew faster and higher than any other airplane ever has; it could, with modifications, have been launched as a space shuttle type in the mid-1960s. An SST built of steel or titanium could have flown at speeds greater

**The needle nose of the Rockwell International B-1 bomber, the long-awaited replacement for the Boeing B-52.**
**OPPOSITE:**
**A happy Air Force F-16 pilot looks up to the camera.**

than Mach 3.0, but optimum engine and aerodynamic efficiency dictated a design speed limit of Mach 2.7.

The American SST was intended to be much larger than its European counterparts, as well, carrying as many as 300 passengers. An enormous and attractive mock-up was created by Boeing, and thousands of engineers were assigned to the project.

As matters unfolded, however, environmentalists expressed concern over potential noise, pollution, and destruction of the atmosphere's ozone layer. Economists pointed out the difficulty of recovering costs in light of soaring fuel prices. The SST became a political hot potato, and the program was canceled in 1971.

163

**LEFT:**
Two Grumman F-14A Tom-Cats in formation.

**OVERLEAF:**
Perhaps no other aircraft has been so in advance of all others for so long as has Kelly Johnson's Lockheed SR-71A, the famous Blackbird.

Yet the American SST—and to a large extent, the Concorde—created a legacy of engineering that is still being drawn upon. From its engineering adventures came a host of advances, including increased use of titanium in structures and significant gains in propulsion technology, that can be seen both in today's civil transports, and in today's military aircraft. The SST led the way in developing "fly-by-wire" techniques that will be standard for the next generation of aircraft; it is also responsible for the use of digital instruments and cathode ray tubes in the modern flight deck. The American SST was intended to make great use of composite structures and metal sandwich construction, and these now appear in everything from light planes to the 767s.

The question sometimes arises as to whether there will ever be an American supersonic transport; the answer is no. But in time there will be a hypersonic transport. The next two or three decades will be spent in working out a solution to the carbon fuel problem and in creating more economical aircraft for its use. The twenty-first century will almost certainly witness the introduction of new fuels (probably methane-based) that permit future transport aircraft to operate in the Mach 4.0-to-5.0 range, reducing the twelve-hour circle to a three-hour circle via the "Orient Express."

The technology and the attitude of cooperation that the development of the SST generated are evident in aircraft development today. Nowhere is this more apparent than in the new designs cropping up in every major country. Sweden is preparing an advanced new fighter to follow the Viggen into service. West Germany, Japan, China, Italy, Brazil, France, and Great Britain are all in the process of developing prototypes in competition for the military and commercial contracts of the future. Sweden and the Soviet Union maintain independent design status, declining to work under formal contract with other nations. Yet no matter how nationalistic and independent each country might like to be, there is the inevitable necessity to introduce foreign technology into each new aircraft design.

# 7. THE DEMAND
## FOR ECONOMY

The cancellation of the American SST was a crushing blow to the pride of the American aeronautics industry. The compelling logic of the past had called for more and more speed, and from this standpoint the SST seemed to suit aviation progress like the capstone of an arch. But the cancellation had a silver lining: the SST would have foundered economically in the brewing tempest of the oil crisis, and most aviation experts now agree that (financially) it was providential that the SST was not completed.

Indeed, the oil crisis gave the industry a chance to demonstrate another aspect of its prowess. Around the world—at McDonnell Douglas, Aerospatiale, British Aerospace, Boeing, and elsewhere—aeronautical engineers immediately determined that the next generation (and perhaps the next several generations) of air transports would have to be oriented toward economy rather than toward additional speed. Of the several designs produced in response to the new priorities, the Boeing 767 illustrates how well the industry did in meeting a new and unexpected crisis.

In aviation, as in art, the relationship of the old to the new is always strong. A superficial comparison of the 1902 Wright glider—one pilot, no passengers, two wood-and-fabric wings, perhaps a 20-mph maximum speed—to the Boeing 767—two pilots, 280 passengers, advanced metal-and-

**OPPOSITE AND FOLLOWING PAGES:**
In the beginning, the pilot was the only "computer" in the cockpit. Information from early instruments had to be integrated in the pilot's mind; the pilot had to respond. As aircraft became more complex, so did instrumentation; but the pilot still translated the data into action. With the modern flight deck, much of the data is used and acted upon by computers, giving the pilot freedom for greater control of the plane.

composite-material construction, and a maximum speed of 625 mph—would seem to discount the possibility that they had any engineering problems in common. And yet they did.

In 1902, a crucial problem facing the Wright brothers was how to design the correct airfoil, or wing curvature and shape, once they had discovered that existing data led to airfoils incapable of providing enough lift.

Three quarters of a century later, the Boeing Company was faced with a problem almost as perplexing. Fuel prices had jumped from 12¢ per gallon in the early 1960s to ten times that, and they seemed sure to continue climbing. A new kind of aircraft had to be created—one that would be far more economical to operate than any had been in the past. To solve this new expression of the Wrights' old problem, the engineers addressed (among many things) the problem of the airfoil.

In some respects, the Wright brothers, working at a point in history when no airfoil had been demonstrated to be correct, had an easier task; they were seeking only enough lift to fly, and the airfoil they settled on had a relatively simple shape and size. If they had chosen a slightly different airfoil, performance probably would not have been significantly affected. The Boeing engineers' problem was far more complex. Hundreds of thousands of airfoils had been designed and tested in the past,

yet none was capable of delivering the lift/drag relationship necessary for the new airplane—the 767—to carry passengers at high speeds at low fuel costs.

In 1902, the Wrights created their own wind tunnel to verify the performance of their airfoils, testing over 200 different shapes. Boeing, which has long had one of the best privately owned wind tunnels in the world, spent more than 30,000 hours testing an aft-loaded airfoil (one whose bulk is loaded two thirds of the way back from the front edge of the wing). Reversing a long trend toward thinner and thinner airfoils designed to reduce drag at high airspeeds, Boeing increased the thickness of the 767 wing profile by 22 percent, compared to the wing thickness of previous Boeing models, and made the airfoil somewhat flatter on the bottom.

The new airfoil made it possible for Boeing to build a longer wing without changing the weight and strength factors, thereby gaining inherent aerodynamic advantages. The thicker wing provided more internal volume for fuel storage, and the positioning of the larger portion of the airfoil toward the rear of the wing allowed the front portion to be made much sharper, thereby reducing drag. Finally, as an almost unexpected bonus, the aft-loaded wing section provided so much additional lift at low speeds that much simpler and lighter high-lift devices (slats and flaps) could be used. Although the new airfoil was hardly revolutionary in appearance, it was of fundamental importance to the new technological quest for greater fuel economy. Still the wing airfoil section that was chosen was but a beginning.

Every airplane has dozens of possible configurations, from high-wing to low-wing, from aft-mounted engines to overwing-mounted engines, from double fuselages to an all-flying wing. Even within a single configuration—for example, a low-wing, pod-mounted engine type—innumerable possibilities exist, each with positive and negative aspects. Here, the Boeing engineers had an immense advantage over the Wrights. In the not-too-distant past, engineers' time limitations barred

**OPPOSITE:**
The Alpha Jet is a joint project of France's Dassault-Breguet and Germany's Dornier companies. It is used in seven different countries both as a combat-crew trainer and as a light strike or reconnaissance aircraft (*top*). The Panavia Tornado is a multinational aircraft built by the British Aerospace Corporation, Germany's Messerschmitt-Bolkow-Blohm (MBB), and the Italian Aeritalia. This "swing-wing" fighter can assume strike, air defense, training, and reconnaissance roles. Top speed is over Mach 2 at its optimum ceiling (*bottom*).

pursuing every possibility; ironically, as the numbers of engineers grew, the actual physical combinations that could be tried diminished because it was impossible to have so many groups interact over a wide variety of types.

## New Tools

With the 767 and others of its generation, the advent of computer-aided design techniques changed the nature of the problem. Now, every member of every design team can maintain up-to-date information on the aircraft's status and (even more important) on each other member's thinking, in real time.

Each team has access to all relevant programs, and changes can be introduced on the various video control displays—which become, in effect, drawing boards. The computer translates the strokes of a designer's lightgun into precise strokes and dimensions. Engineers can brainstorm in a community pattern, without the necessity and expense of creating formal drawings. Computer-aided design thus offsets to some extent the growth in the number of people required to establish a new leading edge.

The unseen effect of a breakthrough design like the 767 lies in the way that the design and construction innovations are passed on to thousands of subcontractors who are all eager to produce components that meet Boeing's exacting standards, even though doing so requires them to adopt new methods of design, measurement, and control. Versions of computer-aided design are required, and some are even linked to the Boeing system. Such innovations are of course applied to the other aspects of the subcontractors' business, as well, benefiting them and their other customers in a spreading ring of improved technology. The linkage is usually hidden, but a well-executed pair of skis, for example, can be traced to a Boeing subcontract for overhead bins, and perhaps even farther to instrument housings on a spacecraft.

Other comparisons can be drawn between the early Wright aircraft and the 767 and its peers. All

have used three-axis controls, and control sensitivity has always been of tremendous importance to their success. To Orville and Wilbur, the degree of sensitivity desired was that enabling them to avoid being dashed into the ground and to avoid throwing the nose so high into the air that the engine stalled. (This formulation is not far-fetched; were it not for the oversensitive elevator controls, Wilbur Wright would have been the first man on earth to fly—not his brother—and we would today celebrate December 14 as the birthday of flying. Wilbur pulled back just a little more than he should have on the stick that controlled the elevator, and the Kitty Hawk Flyer leaped to the height of perhaps 15 feet before stalling and dropping to the ground. To the purist scientist residing in the Wrights, that was not enough to be called a genuine flight.) In the 767, control sensitivity translates into fuel economy and is governed by the flight management system (used in both the 757 and the 767).

Amidst all the changes surrounding the replacement of piston-engine Connies by jet 707s, one thing that changed very little was the flight deck. Pilots were still required to integrate the intelligence obtained from myriad readings of individual displays, and to translate this into control adjustments. It was an excellent system for its day, but the pilots needed time to make the translations mentally. The method also left room for differences in interpretation, based on pilot skill and experience, and this was not in the best interests of safety. The significance of an interpretation varied with the situation: a pilot flying on a clear day at 30,000 feet, with everything running smoothly, can absorb information readily; a pilot making a landing approach in a thunderstorm, with one engine shut down, a cabin full of frightened passengers, and the radios acting up, faces quite a different situation and needs all available help.

The space age brought digital electronics to the jet transport, providing the same degree of precision that spacecraft enjoyed on their journeys to the planets. Boeing created the flight management system (FMS) specifically for the 757 and 767 aircraft. The FMS provides an exactitude of

**RIGHT:**
The design considerations of the Boeing 767 extended even to the brake assemblies shown here (*top*). This is the interior of the French Airbus, a plane that offers stiff competition to American models (*center*). The integrated flight deck has permitted the reduction of the flight crew from three to two, eliminating the position of flight engineer (*bottom*).

measurement that is unmindful of potentially distracting vibration, temperature, weather, distance, time, or emotional condition, and it transmits information to the pilots through multicolor displays that permit instantaneous and consistent interpretation. The instruments always provide the pertinent information on the systems, the weather, and navigation; consequently, the pilots are not required to digest and analyze disparate information.

Three concepts underlie the FMS: simplification, redundancy, and automation. In one typical function, the system ensures maximum fuel economy by linking together the autopilot, the automatic throttle system, and the navigation systems. The pilot makes manual adjustments during the actual take-off. Almost immediately afterward, however, the autopilot is engaged, and the FMS controls the airplane with a precision that is impossible for a human to match. If a failure occurs, a backup system is automatically engaged, obviating the former need for hurried cooperation between pilot and copilot reading emergency checklists. The pilots are informed of what has happened, but the malfunction is routinely overcome automatically.

This might suggest that the pilots have surrendered some control; in fact, the opposite is true. They now have time to bring their experience and judgment into play without being distracted by procedural mechanics. When the cured problem is announced to them, they can analyze why it happened, and what additional steps might be desirable. The system is so efficient that a two-person crew (rather than a three person crew, as before) is considered entirely adequate.

A side benefit of the precision of the FMS is the passenger comfort it provides. Corrections are introduced so subtly that the passengers—and the pilots—are unaware that they are being made. The result is a soft ride, even in turbulence.

## New Materials

Like the Wrights, the Boeing engineers were concerned about weight. The Wrights had so scrupulously calculated their requirements that they made

Films of a series of tests conducted by NASA revealed a sequence of structural deformation; this information was invaluable for the design of new cabins.

the right wing 4 inches longer than the left to compensate for the difference between the weights of the engine (180 pounds) and of the pilot (140 or 150 pounds, depending upon whether it was Wilbur or Orville). The total gross weight of the Kitty Hawk Flyer was about 760 pounds. The gross weight of the 767 is 300,000 pounds; yet the need to save weight is equally acute.

The effort to create an economics-based leading edge in flight implicated a whole series of considerations, not themselves technically at the leading edge, that nonetheless contributed substantially to the 767's being there. This is particularly true with respect to reducing the aircraft's weight. The space age lent a hand in weight control, too. Boeing shaved some 6,500 pounds from the 767 by using new materials. Boeing engineers had tested most of these well in advance by using them on smaller components of aircraft actually in service, where advanced composite materials were found to be lighter, stronger, and more corrosion-resistant than the metals they replaced.

The very first metal airplanes were made of steel, with all the disadvantages of weight and rust that steel implies. By the early 1920s, aluminum became the preferred metal, and over the next sixty-five years, continuous progress was made in improving aluminum's strength and durability. Yet some problems are inherent in the use of metal, including failure from fatigue induced by flexing and by inevitable corrosion.

The space age alternative seems to be to use carbon fiber–reinforced plastic, also known as graphite/epoxy material, in parts requiring stiffness or high strength—exactly as the Wrights used ash instead of pine. Typical applications occur at places where tension or compression might be encountered, such as in the ailerons, the spoilers, and the rudder. A Kevlar/epoxy finish is used to coat areas that might be struck by debris thrown back from the ground or bumped by ground service vehicles. The Kevlar/epoxy is used as a structural material on items not subject to high stress, such as overhead storage bins. Composite materials also serve economy because of their smooth

finish; being glasslike, they require less maintenance and reduce drag.

The world has been quick to respond to the possibilities for better speed and economy that the new materials provide. In Great Britain, British Aerospace, Inc., is producing a host of new designs ranging from the Experimental Aircraft Programme (EAP), an advanced fighter, to the Advanced Turbo Prop commercial transport. In France, the remarkable Avions Marcel Dassault-Breguet Aviation firm has put forward the Rafale.

The EAP and the Rafale feature twin engines, canard surfaces, and air intakes, which are continuously modified as the designs progress. Both feature the use of composite materials, advanced alloys, new manufacturing techniques, and advanced flight controls. The computer-aided designs optimize the size of the aircraft for multiple roles without enormous manufacturing expense.

Smaller nations also strive to maintain their place in the forefront of aircraft development. Israel Aircraft Industries, Ltd., began over thirty years ago in an antiquated overhaul facility; it now has annual sales of over $1 billion. In 1986 it produced the LAVI. In keeping with the ''born in combat'' mentality of the Israeli Air Force, the LAVI is the logical next step for an industry that began by modifying World War II fighters and, over time, established itself as a worldwide leader in the manufacture of combat planes. Economics or politics may interfere with production of the LAVI, but the flight of the prototype provides convincing evidence of the plane's capabilities.

## THE FINAL LINK

Ultimately, of course, the power and economy achievable by the Kitty Hawk Flyer, the 767, or any other aircraft depend upon the engine.

Since the beginning of the jet age, the progress of engine thrust, engine reliability, and fuel economy has been staggering. The engines for the Messerschmitt Me 262 generated a little less than 2,000 pounds of thrust and had a life expectancy of twenty-five hours. The engines on the Concorde

develop more than twenty times the thrust and have hundreds of times the reliability. The General Electric engines designed for the SST were capable of more than 60,000 pounds of thrust.

But today, sheer power is not the essential requirement of jet engines. During the piston-engine era, aircraft were almost always considered underpowered; now ample power exists to ram fighters to Mach 2 and beyond, but the need to do so is not always present. Other factors (particularly economy) enter in, and among these is the ability to function under the stress of high G loads—the effective multiples of gravity induced by maneuvers.

Enormous improvements have been made in the materials that go into turbine engines and in the lubricating systems of these engines. Once, a twenty-five-hour engine life was considered satisfactory. Now, jet engines can run for thousands of hours before requiring replacement. For some jet engines, the danger of corrosion due to lengthy exposure to the elements is a greater risk than that of simply wearing out.

The General Electric J-47 engines that powered the technologically influential Boeing B-47 were conventional axial-flow turbo-jets of about 5,000 pounds of thrust, and they had a specific fuel consumption of about 1.2 in cruising flight (specific fuel consumption, a measure of jet engine economy, refers to the weight of fuel required to achieve a given power output; the lower the specific fuel consumption number, the more efficient an engine is).

The next major advance came with the Pratt & Whitney J-57 engine, which in its JT-3 form made the commercial jet liner possible. The J-57 was used on the B-52 and a host of other aircraft; it had a takeoff thrust of about 10,000 pounds without water injection, and a takeoff thrust of as much as 13,750 pounds with the system. Specific fuel consumption with this engine improved to the .8 level, and noise levels declined.

Over the decades, the state of the engine industry changed vastly. In the piston-engine field, the venerable Curtiss Wright Corporation had been Pratt & Whitney's primary competitor. After World War II, a host of firms—Allison, Westinghouse,

Lycoming, and others—tried to compete for contracts, but by the late 1950s, only two firms were seriously competing for contracts on advanced engine types: Pratt & Whitney, and General Electric. Much of the latter firm's success was due to Gerhard Neuman, an intense, dynamic man, who created the variable stator system on the General Electric J-79 engine. This engine was in production for more than twenty-five years and powered many of the free world's fighters and bombers. With afterburning, it had a thrust of almost 18,000 pounds and a specific fuel consumption of 1.97.

The next major advance in engine design was the turbo-fan, which added a large-diameter fan element to the engine (in effect, a propeller pushing masses of cold air through the engine) and obtained greater power with improved fuel economy—one of the few instances of an advance that seemingly offered something for nothing. The B-52H used Pratt & Whitney TF33 engines, which were rated at 17,000 pounds of thrust and attained a specific fuel consumption of .56 at cruise.

If a large-diameter fan improved power and fuel

**The F-15 Eagle—carrier for one of the most controversial weapons of the day, the anti-satellite missile—on a low-level pass near Las Vegas.**

consumption, an even larger-diameter fan should do better still. General Electric used this reasoning to create the CF-6, while Pratt & Whitney delivered the JT9D. These high-bypass-ratio turbines deliver up to 48,000 pounds of thrust and have a low specific fuel consumption number. The combination of economy and reliability offered by these engines has made it possible for such twin-engine aircraft as the 767 to be declared safe for transoceanic passenger flights.

The enormous power and fuel-efficiency requirements of the 767 and its military and civilian contemporaries can be met only through the use of modern computer devices. Once again, the aerospace industry has witnessed a push-pull of need and circumstance. Modern computing machines and techniques are changing the industry even more drastically than did the introduction of the jet engine, and their influence is felt throughout the world.

Computer-aided design (CAD) has eliminated acres of bullpens, filled with hundreds of engineers hidden by eyeshades and working away with their slide rules at drafting boards. In their place, minicomputer-driven work stations equipped with high-resolution color graphics have proliferated. Testing, too, has been revolutionized. In 1978, by means of the Cray supercomputer, it became possible to perform mathematical simulations of the flight performance of prototype designs *before* the actual machines are flown.

The CAD and supercomputing processes are complemented by computer-assisted manufacturing (CAM), which links the designer's work station to machine tools on the factory floor and to subcontractors. Such manufacturing areas are often robotized to achieve even greater precision and economy.

The same types of computers have been adapted to create simulators for training. The outstanding successes of NASA's space program would have been unattainable if the crews had not received the quality of simulator training that computers alone can provide. Computers are even more conspicuous in their roles in ground control and navigation of flights, where techniques of inertial guidance are combined with ancient methods of navigating by the stars.

## A LOOK AT TOMORROW

Perhaps the best current example of the distillation of computer design and control techniques is the space shuttle, which would have been impossible to realize without use of every one of the tools described heretofore. The immense National Aeronautics and Space Administration team, with its headquarters in Washington and its massive centers in Houston, Cape Canaveral, Sunnyvale, and elsewhere, could not have coordinated the efforts of the space shuttle program without advanced communications methods and computers. The thousands of contributing firms—from giant Rockwell International, which fabricated the orbiters, down to anonymous suppliers of the small plastic bags that dispense M&M candies to the shuttle crews—could not have worked together without taking full advantage of today's expanding combina-

**RIGHT:**
The Boeing 747 was never designed to carry the space shuttle, but when the need emerged the Boeing engineers tackled it with success.

**OVERLEAF:**
No matter how complicated or sophisticated aircraft become, no matter how exotic the computers, the ultimate decisions will always be lonely ones, the responsibility of the pilot in command. Here the crew of a Canadian CF-101 strap in, their thoughts on the mission ahead.

tion of modern engineering and communications.

And with all of this, the space shuttle is itself a flying truck—a transporter designed to place the really sophisticated elements of science into orbit. The space shuttle is best seen as a facilitator of the leading edges of tomorrow, which range from a gigantic space telescope (possessing the capacity to peer 14 billion light-years away to the very edge of the expanding universe), to the possibility of a space-derived cure for cancer to the construction of a permanent space station.

These gigantic leaps into the future, which seem inevitable to us now, would have exceeded the powers of the Wright brothers to imagine. Yet they are merely steps along the way from the wonderful first flight at Kitty Hawk to a future even now beyond the farthest range of our imaginations.

# 8. LEADING NOWHERE

In the eighty years of flight, millions of aircraft and tens of thousands of designs have been produced. Not surprisingly, some of the designs were dead ends despite embodying crucial elements of a leading edge of flight.

The failures have most often been represented in a few comic films—excerpts from newsreels, primarily—that have cast an altogether unfair light on the situation. The film chiefly responsible for this, *Aeronautical Oddities,* was put together in the 1930s and has since been pirated on every occasion when footage of a "funny" side of flight was needed. It contains the all-too-familiar scenes of a bird-man jumping from a bridge, of another with a similar set of wings and tail leaping futilely from a rock, and of a venetian blind–winged aircraft

The Dornier DO-X, a twelve-engine behemoth that was intended to revolutionize air transportation, to be a true "ship of the sky."

**OPPOSITE:**
The Cessna floatplane is everything the DO-X was not: small, practical, and enormous fun to fly.

being towed and then suddenly collapsing in a heap.

Yet most of the leading edges that ultimately led nowhere were founded not on eccentricity or foolishness, but on wild, undeveloped glimpses of the future that precluded taking the present into account. Often, an inventor fastened onto an idea that was inherently good and made aerodynamic sense but was ahead of its time, resulting in a system or procedure that worked perfectly in an aircraft that did not.

## THE ROMANCE OF FLYING BOATS

A leading edge might bring a whole discipline to a fine point, only to have other aeronautical

advances relegate it to the dustbin of history. A classic example of this is the creation of aircraft capable of taking off and landing on water. Henri Fabre made a tentative flight in a seaplane at Martigues, near Marseilles, on March 28, 1910. On January 26, 1911, near San Diego, Glenn Curtiss flew the first practical seaplane—essentially a standard Curtiss pusher biplane mounted on an absurd-looking main float (6 feet wide by 5 feet long) that was preceded first by a nose float and then by a hydrofoil. The combination worked, but its bulkiness and complexity were clearly unsatisfactory. In short order, the float apparatus was modified into a single boat-shaped main float. Then, in July 1912, Curtiss launched what many consider to be the most romantic period in aviation, with the world's first flying boat, ''The Flying Fish.''

**Howard Hughes built the largest airplane ever, the H-4 flying boat with a 320-foot wingspan. It flew only once, for about a mile, and ultimately became a museum piece.**

Seaplanes (conventional-looking aircraft with single or double pontoons for flotation) and flying boats (shiplike hulls with wings) had a tremendous appeal in the days before many landing fields existed, since most could be tied up at ordinary boat docks. Each type offered advantages for specialized uses: the seaplane was adaptable for use as a scout plane on naval vessels, where it could be launched from catapults and recovered by cranes; the flying boat was used for standard military purposes and for carrying passengers relatively safely on overwater flights, at a time when engines were not reliable. When wheels were fitted to either seaplane or flying boat, the result was quite naturally known as an amphibian.

The flying boat richly portrayed the glamour of flight and for forty years served aviation well in

every part of the globe. World War I gave great impetus to the flying boat's development, with Curtiss equipping the navies of both the United States and Britain with large wooden biplanes for patrol and antisubmarine work. The H-16 was a large twin-engine biplane, and the HS-2L was a scaled-down single-engine version—both using the 400-horsepower Liberty engine. The H-16 was developed by the British into the Felixstowe series of aircraft, used by both navies for many years.

The first aircraft to cross the Atlantic (from May 16 to May 27, 1919, with three stops) was the Curtiss NC-4, a four-engine development of the World War I flying boats. From this point until the early 1930s, flying boat development closely paralleled land plane development. Flying boats of basically similar configurations were developed in England,

**The HK-1 was built entirely of a material called "Duramold," wood impregnated with plastic and molded in the exact size and shape to handle calculated stress.**

France, Italy, Japan, and the United States. Germany was host to an advanced trend toward all-metal monoplane flying boats, whose performance dominated the second decade of the century.

The era of Empire boats and Clipper ships—the American and English flying boats that linked widely separated national interests—is perhaps unsurpassed in its romance and luxury. The very best work of the time emerged from the genius of Juan Trippe, in his pioneering efforts with Pan American Airways. Trippe's prescient demand for safe, profitable, passenger-carrying flying boats pointed out a solution to the worldwide problem of long, overwater routes and insufficient airfields for land planes. At any seaport or lake, the big flying boats could alight in the water, taxi up to a fuel drum—supported pier, and deliver passengers and mail.

Trippe called on the three major American manufacturers of flying boats to produce a totally new breed of aircraft: a flying boat that could carry passengers across either the Atlantic or the Pacific. The manufacturers responded with four brilliant designs whose efficiency for a brief period exceeded that of any land planes.

The first of these was the Sikorsky S-40, a rather ungainly four-engine airplane that entered service in November 1931. The S-40 could carry forty passengers at 100 mph for 800 miles and was highly successful. It was succeeded in 1934 by the brilliant S-42, a sleek, high-wing, four-engine aircraft with cowled engines, variable-pitch propellers, wing flaps, and a speed of 140 mph with thirty-two passengers. At 43,000 pounds, its gross weight was more than twice that of the biggest land-based airliner of the day, the Douglas DC-2 (which carried only fourteen passengers at the higher speed of 170 mph).

The Martin M-130 flying boat came next. A beautiful aircraft, the M-130 incorporated all of the S-42's advances in a larger, more powerful package. It made the first Pacific crossing (from San Francisco to Manila) in November 1935, flying via Honolulu, Wake Island, Midway, and Guam.

The Sikorskys and Martins—captured in our memories forever as China Clippers, although they bore many other equally exotic names—were at the very forefront of aeronautical development in efficiency and safety. They combined all the leading edges of land planes with high-aspect-ratio wings that enabled them almost to match land-plane speeds while exceeding the distance and passenger-carrying capabilities of land planes. The progression reached its peak with the lovely, triple-tailed Boeing 314, the deep-chested flying boat that carried so many Allied leaders on secret missions during World War II. The 314 was as far ahead of other flying boats of the period as the B-29 was in advance of contemporaneous bombers. Powered by four 1,550-horsepower Wright Cyclone engines, the 314 could carry seventy passengers at 145 mph for 2,400 miles. Only twelve were built, and none exist today.

The great flying boats were so firmly established in the psyche of aviators that many did not notice that World War II had created the basis for their demise. All over the world, great airfields with long runways, efficient communications, and secure maintenance buildings had sprung up. The facilities that had been unaffordable in peacetime were built during the war years with profligate disregard for expense; and as they were built, they removed the raison d'être for the flying boat. Because the big boats had been so useful for so many years, proponents imagined that their services were indispensable. A parallel case involved railroad executives who plowed enormous sums of money into new passenger trains after World War II, only to see automobiles and airplanes capture their market.

Two spectacular efforts, both couched in the same insular detachment from reality, produced beautiful airplanes destined for failure. One of the two, the English Saro Princess, was broken up after years of storage, an embarrassment to the manufacturer and to the government. Its American counterpart, the magnificently outrageous eight-engine Hughes Flying Boat, is still the largest aircraft ever built. It is preserved today in Long Beach, California, nestled in close proximity to another, more productive giant, the Queen Mary. The Flying Boat, its piano-quality wooden splendor intact, is still capable of producing awe and still echos the dark, brooding eccentricity of its sponsor.

Technically the Saro was the more sophisticated of the two flying boats, being all metal, weighing 345,000 pounds, and drawing power from ten Proteus turbo-props. The entire program was put together on faith, with no preliminary survey of the market to determine the need; alas, no need existed, and the first prototype—together with the airframes of the second and third—was cocooned.

The Hughes story represents the classic case of the man with tunnel vision. The tunnel here was enormous, and the vision became overladen with hubris. Like Hughes's remarkable racer, his flying boat was an exercise in engineering integrity. It

stemmed from two wildly volatile, standards-be-damned individuals, Hughes and Colonel Virginius Clark. Clark was in the Hughes/J. V. Martin mold: utterly irrepressible, irascible, disdainful of convention, difficult to like, impossible not to admire.

Clark invented the Duramold process used in the Cellini-like construction of the big flying boat. Duramold was a method of plasticizing wood and molding it so that, like modern composite materials, the part could be matched perfectly to the strength required. The skeleton of the Hughes boat is composed of a series of beautiful Duramold structures, each arching to meet the stress.

But the Saro Princess and the Hughes HK-1 faced the same problem: time had passed them by. Flying boats were no longer required, no matter how big, how fast, and how beautiful, and no matter how great the nostalgia of the designers.

## UNTIMELY BREAKTHROUGHS/
## TIMELESS FLAWS

Other leading edges came to grief simply because the aviation world was not prepared for them. An example is the Lawson Airliner of 1919. Alfred Lawson, a man of religious zeal as well as aeronautical insight, conceived of a passenger airliner before airlines, airways, airfields, air terminals, and (perhaps most fundamentally) airline passengers existed. On its trial flight, on August 28,

1919, Lawson took it on a record-setting trip to Chicago, Toledo, Cleveland, New York, and Washington. He demonstrated to the world the functional capability of an airliner by carrying fourteen persons from New York to Washington, D.C., in four hours and twenty minutes, against strong headwinds. The aircraft was fully competitive with anything in the air, but it offered too much too soon, and so sank into oblivion.

Many other leading edges also missed the brass ring. The J. V. Martin Kitten, a remarkable little biplane with genuine retractable landing gear (hindered only by the fact that it could not fly) has already been mentioned. The Kitten was but one design produced by J. V. Martin, a small, gutsy,

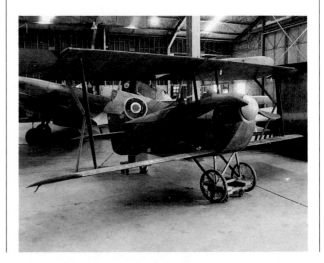

190

volatile man whose red hair and sometimes belligerent demeanor might have earned him the nickname Seminole Sam in another generation. Martin was a ship's captain, a pioneer aviator, an inventor, a manufacturer, and (repeatedly) a litigant.

Martin became a gadfly to the U.S. Army Air Service, hooking the generals with offers of a "free" airplane, and then badgering them for contracts. He benefited most from the association by the prestige it gave him with Congress; it was a rare congressional hearing on the Air Service in the immediate postwar years that did not have J. V. Martin as an expert (and caustic) witness. After World War I, he instituted suits totaling $96,000,000 against a host of competitors, alleging patent infringements of his many inventions. Some of his inventions, such as the retractable landing gear, were excellent; most, however, were terrible and indefensible by any normal engineering logic.

Besides the earthbound Kitten, he created for the War Department the J. V. Martin Cruising Bomber, a handsome biplane exquisitely fashioned of the highest-quality material and embodying most of Martin's loony-tune inventions. The Cruising Bomber gave form to several ideas that were to haunt engineers for years and even reappeared in some later failed leading edges—including submerged engines designed to drive the propellers by shafting, and wing-tip ailerons. The plane never flew because the transmission system failed the Air Service tests, and much to Martin's chagrin it was ultimately used as a target for gunfire tests.

As an aside, it is interesting to note that Martin was the man who suggested to Winston Churchill the concept of a giant, airborne, twin-hulled, ocean-spanning freight carrier; Churchill passed this on to the British admiralty, who contacted Henry Kaiser, who contacted Howard Hughes. The first paper designs for what became the Hughes flying boat were twin-hull designs, and thus the small circle of eccentrics, J. V. Martin, Virginius Clark, and Howard Hughes, was neatly circumscribed.

Across the ocean lived a controversial contemporary, Tony Fokker, whose remarkable career lasted from 1912 until his premature death in 1939. Something of an eccentric himself—he constantly fed himself spoonfuls of granulated sugar—Fokker had combined great piloting skills with great pirating skills. He took his ideas where he found them, and after designing a series of aircraft of varying quality, achieved some success with the Fokker Dr I triplane. This highly maneuverable aircraft, perhaps the nearest equivalent to a Sopwith Camel, was inspired by the Sopwith triplane. The Fokker Dr I is still the one type most evocative of his name; and it was in this plane that the Red Baron was killed.

Considering that if three wings were good, more would be better, Fokker ordered his designers to build a quintaplane, with three wings forward and two farther aft, in tandem. The engineers protested but Fokker overruled them. One squirrelly flight convinced him of his folly, and the quintaplane was quietly disassembled and abandoned.

Designers' fascination with variations in the size, shape, number, and configuration of wings has persisted to this day. Count Gianni Caproni had enormous success with biplane and triplane multi-engine bombers and a penchant for experimentation; these led his firm to make a long line of successful conventional types and also some outlandish failures. The first of his wild nonflyers was the Caproni Ca 60 Transaero, a nine-wing, eight-engine flying boat intended to carry 100 passengers across the Atlantic. This exercise in gigantism

**RIGHT:**
The Sopwith Triplane inspired a designing frenzy of multi-wing aircraft, almost none of which were equally successful.

made only one takeoff, an uncontrolled hop to 60 feet in altitude that was immediately followed by a nose-down crash into the sea.

A few years later (at the direction of the Italian Air Ministry), Caproni created the least attractive aircraft of all time: the Caproni Stipa. The Stipa featured an engine mounted within a mammoth barrel fuselage that was supposed to generate thrust by means of a venturi effect. This aerial cartoon, resembling Porky Pig doing an impression of a wind tunnel, was marginally successful—that is, it did not crash. Despite the Stipa's unimpressive performance, the French firm of Les Mureaux acquired a license for the Stipa patents and planned to build a twin-engine version, but this monstrosity never materialized.

Caproni's last fling at the unnatural came with the Caproni-Campini of 1940, claimed at the time to be the first jet-propelled aircraft to fly. The

**Howard Hughes wanted to sell the FX-11 to the Air Force as a photo-reconnaissance aircraft, but his insistence on doing his own test flying led to a near-fatal crash. Although a second plane was flown successfully, no orders ensued.**

Caproni Campini followed the general idea of the 1910 Coanda in using a piston engine to drive a variable-pitch, ducted fan compressor. A ring of fuel injectors located behind the compressors ignited fuel to provide jet thrust. On its first cross-country flight, it averaged only 130 mph—a performance that had the advantage, at least, of not initiating jet lag.

## BIG, AS IN BRONTOSAURUS

The gigantism that afflicted Count Caproni's creations had similar effect on one design of the normally cautious Claude Dornier. Dornier's firm made a series of wonderfully efficient flying boats, including the famous Wal in which so many records were set. But in 1926, the idea of creating a flying ship intoxicated him, and he poured his resources and energy into creating a giant twelve-engine

flying boat with a wingspan of 157 feet and a 100-passenger carrying capacity.

The airplane, which emerged in 1929, was attractive, but it was far too heavy and drag-producing for its power. It made a single, ago-nizing, ten-month-long flight across the Atlantic, beset by every sort of humiliation—from being unable to get airborne at the Canary Islands to being forced down at sea and having to taxi 60 miles to port to suffering a fuel tank fire that almost burned off a wing. Mussolini was nonetheless im-pressed, and ordered two for Italy. In the end, these were never used except for test purposes, and the original DO-X ended its life in a Berlin museum,

An unusual—and success-ful—giant leading edge was the conversion of sur-plus Boeing 377 Strato-cruisers into Super Guppy aircraft.
**OPPOSITE:**
MBB has developed a family of helicopters, ranging from this multi-purpose Bo 105 to the sophisticated PAH-2 anti-tank.

where it was destroyed during World War II.

Another example of an aircraft with a hyper-thyroid condition was the Tupelov ANT 20, known as the Maxim Gorki, at the time the largest land plane ever built. Its wingspan was 206 feet; it weighed 84,000 pounds fully loaded; and it sup-posedly cost 6,000,000 rubles (or about $20 mil-lion today). Eight engines pushed it along at little over 100 mph. The Maxim Gorki carried a printing plant, a photographic studio, a radio station, and a movie theater, as well as a sign for displaying slogans under the wings. It flew for a year before colliding with an escorting fighter during a propa-ganda flight over Moscow.

## THE DO-EVERYTHING TRAP

Some flawed leading edges were the result of attempting too much too soon. One of the most notorious aircraft of all time, the German Heinkel He 177 bomber, was the victim of an attempt to put 5 pounds of ingenuity into a 2-pound sack of possibility. Officially, the Luftwaffe called the He 177 *Greif* (for *griffin*), falsely but appropriately suggesting *grief* as a cognate. Informally, the crews called it the *Luftwaffenfeuerzeug* (or *Luftwaffe petrol lighter*), for its tendency to catch fire in the air.

In many respects, the airplane was among the most advanced in the world, almost identical in size to the Boeing B-17G Flying Fortress and having similar performance capabilities. The faults in the design resulted from the idiotic specifications imposed by headquarters. The German Air Ministry demanded far more than any company could have delivered, and the He 177 had to be made inordinately complex in order to approach the unreasonable mission requirements.

The first and most basic error was designing the airplane to have four engines located in two nacelles. Essentially, two engines were coupled to create one larger powerplant, driving a single propeller. The idea made sense aerodynamically and structurally, but it was total nonsense from the point of view of maintenance and safety. Then, in a masterpiece of Goeringesque madness, the Air Ministry issued a requirement that the He 177 become a Stuka, able to dive-bomb. This meant that the entire aircraft had to be strengthened immensely, with a corresponding growth in weight and reduction in performance. The absurdity of the command was comparable to insisting that the B-17 become a dive bomber; it was not unlike requiring a cement truck to be able to drag race.

One He 177 after another crashed, either in test or on operations; Hitler thereupon ordered an acceleration in production. In 1944, with Europe in flames, almost 600 of the big bombers were completed, and some participated in the forgotten "little Blitz" of England in the spring of that year. By mid-1944, fuel was in such short supply that most of

The Herrick Converta-plane, an idea overtaken by events.

In the hands of skilled test pilot James "Skeets" Coleman, the Convair XFY-1 "Pogo" achieved true vertical liftoff, but it was not practical for ordinary pilots.

the He 177s were grounded on airfields throughout Germany to provide target practice for marauding American fighters. Far more He 177s were destroyed by strafing than were lost on missions.

## UP, UP, AND THEN AWAY

The idea of a vertically rising aircraft has fascinated the human imagination for centuries: Leonardo's sketches reveal his grasp of the basic concept—a screw mounting vertically through the air. The problems of vertical flight were more formidable than those faced by the Wrights, however, and not until the 1930s did progress begin to be made, culminating in the successful Sikorsky developments during the war.

Almost every possible variation in configuration has been tried for vertical flight, from rotors to tilt wings to tilt engines to direct-lift jet engines. One of the most successful attempts prior to the helicopter was Gerald Herrick's Convertaplane. Wishing to improve on the autogyro, and deciding that helicopter technology was not yet attainable, Herrick created an intermediate step that might have been successfully marketed if Sikorsky's efforts had failed. The Convertaplane looked like a conventional biplane, with a rather smaller upper wing mounted on a stalky pylon. The upper wing was actually a rigid, two-bladed rotor that converted to a single cantilever wing in flight. This combined the advantage of the autogyro's vertical lift capability with a higher cruising speed.

Although the Convertaplane worked, it was not a resounding success. Unexplained drag problems (never solved) reduced performance in the horizontal regime. And like many other inventors of the time, Herrick faced financial problems that led him to abandon his efforts. Later inventors played variations on his theme to produce various tilt-wing and tilt-engine aircraft; all of these failed because they had inadequate power to combine the two flight regimes. Not until Bell Aircraft introduced the XV-15 —the product of more than twenty years of experimentation and hundreds of millions of dollars—did Herrick's dream come true.

## WINGS AWRY

From the days of the Wrights and Langley, the desire to improve performance through a variation in wing shape or configuration has never varied. Some inventors were attracted by the symmetry of a basically round wing: it just looked right. Over the years, perhaps a dozen different aircraft have used some variation of the low-aspect-ratio (relationship of length to breadth) round wing, counting on the effect of the propeller blast over the entire wing to offset its inherent aerodynamic inefficiency in comparison to a wide, slender, high-aspect-ratio wing.

Vought Sikorsky seemed to get closest to success with its V-173 Flying Pancake, which actually flew pretty well. The design became the basis for an advanced fighter designed at the end of World War II but scrapped prior to flight because it had been designed with piston rather than jet engines.

Other inventors wanted to have big wings on takeoff and little wings in flight, and achieved this in a variety of ways. In France, M. Makhonine, a Russian expatriate, created an airplane that took off with its wings extended to 69 feet, and then reduced its wingspan in flight to 43 feet by stowing it internally with a roller system. The system worked but did not offer sufficient advantages over conventional flaps to offset its weight and complications. The elegance of the conception was not matched by cleverness of execution.

The Russian Nikitin-Sevechenko IS-1 was even more ambitious. On the ground, it was a conventional biplane, possessing a biplane's characteristic suitability for short-field use. On takeoff, however, the pilot retracted the wheels and the lower wing, which folded up neatly into the gull upper wing. The basic idea was workable and—except for some peculiar airflow problems when the wings began to be merged into a single surface—involved no insuperable difficulties. The gain in performance over that of conventional aircraft was modest, however, and no production of this model ensued. The Soviet engineers undoubtedly realized that the complicated device's massive hydraulic system

The Custer Channel Wing concept almost entered production on two occasions, but fate intervened both times.

**BELOW:**
The Vought XF5U-1, based on the "Flying Pancake," was almost ready for flight in 1946 when the Navy cancelled the program.

would have fared ill in the freezing Russian winters.

In America, a far simpler design was tried to achieve short-field performance: the Custer Channel Wing. This design was the brainchild of Willard R. Custer, who once saw a tornado lift the roof off a barn and had an instant insight into a new aeronautical phenomenon. An airplane flew by moving its wing through the air, but the barn roof had flown when the air moved over it. It was clear to Custer that if he could construct a wing in such a way that air could be moved over it, the airplane could fly independent of forward speed.

He turned to Bernoulli's principle and built a series of airplanes with great half-barrel-shaped airfoils. The engine and propeller were mounted within the half barrel; and when they sucked air over the wing, the airplane did in fact fly. But his inventions did not gain acceptance, partly because their role was somewhat filled by helicopters and partly because Custer believed so strongly that his Channel Wings could do everything better—high speed, low speed, personal plane, transport—that he did not receive the financial backing he might have gotten if he had concentrated on a single mission.

## FLYING WINGS

While many designers tried to vary the shape of the wing, some tried to make the wing the entire airplane. Throughout the history of flight, certain visionaries had seen the empennage (the rudder and elevator assembly maintained at the rear) as mere baggage, contributing weight and drag but little else—if only control could be established without them. Great successes with the tailless, all-wing formula were achieved almost from the start. Lieutenant J. W. Dunne designed a tailless pusher that flew very well in 1911, and this was followed through the years by such descendants as the Simplex-Arnoux tailless racer; professor G. T. R. Hill's series of Pterodactyls for Westland; the ingenious designs of the Horten brothers in Germany; and the deadly Messerschmitt Me 163 rocket fighters, derived from the work of Alexander Lippisch.

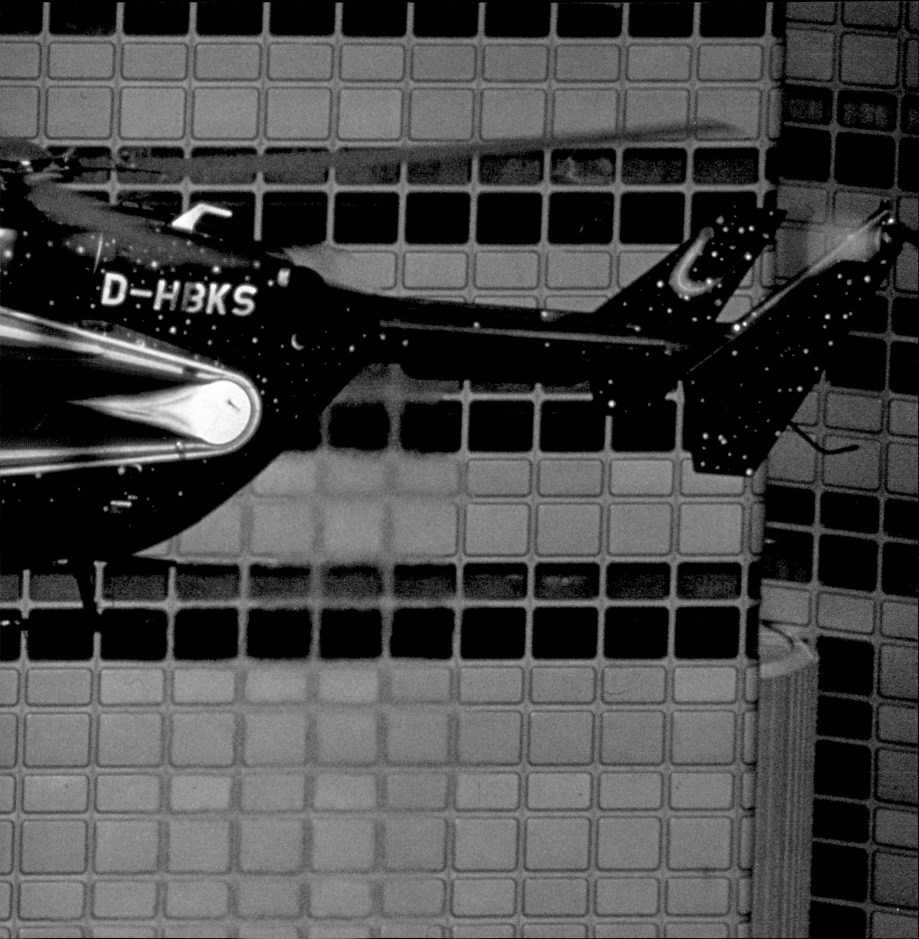

The Me 163s were the only flying wings ever built in quantity, and they were extraordinarily fast, achieving more than 620 mph as early as 1941. They flew beautifully, but their rocket engines made them more dangerous to their pilots than to the enemy.

Of all the variants of the flying wing, the ones most romanticized (and perhaps most criticized) were those produced by John Northrop. Northrop, who was the genius behind the Lockheed Vega and the Northrop Alpha, among many other designs, was determined to make the flying wing the plane of the future. He designed a series of experimental types before receiving orders from the Army Air Force for the XB-35, a huge, piston-powered flying-wing bomber. It arrived just as jet engines did, and the aircraft was subsequently modified to become the jet-powered YB-49.

The big Northrop flying wings were not successful, primarily because they required a stability-augmentation system that was not yet available; but although they passed unused into history, they remain a source of heated speculation. One rumor, which must please the ghost of the late John Northrop, is that the new stealth bomber from the firm he founded will be a flying wing.

The flying wing concept failed to achieve production in most instances because the advantages

**PRECEDING PAGES:**
The MBB BK 117 twin-engine helicopter was developed jointly with Kawasaki of Japan and has been put to both civil and military uses.

**BELOW:**
One of the most romantic and persistent of the ''might-have-beens'' is Jack Northrop's Flying Wing.

it offered were offset by difficulties in control and stability. Almost certainly, many of the flying wings could have been placed in service and done well. As with so many other great ideas, the need for the airframe and the availability of money to perfect it never coincided.

## MAVERICK WINGS

The quest for improved aircraft shapes continues. Perhaps the most promising and the most poignant attempt of recent times is the Lear Fan, the final product of Bill Lear's wild genius.

Lear had spent his life making millions of dollars by doing things that people told him could not be done. Against all advice and despite active, hostile opposition, he pioneered the executive jet revolution with the classic Learjet. The mechanical basis for the Learjet was the airframe of a Swiss fighter plane that had not been put into production. Lear determined to build an airplane that would be low in drag, so he created a small-diameter (fighter-plane-size) fuselage, in which the passengers could sit comfortably but could not walk upright. Everyone told him that no one would buy such a plane. He created an airplane that was sold complete, instead of just building a shell that buyers could then custom-outfit with their own interiors

and avionics. Again, he was told that this was the wrong thing to do.

But Lear saw what they did not: that there was no other airplane on the market that could compete in speed with the airlines, and certainly nothing on the market with airline speed that could operate out of small airports. The Learjet went into the history books as Lear's greatest success. Today, it remains the most identifiable of corporate jets, and its many successors (from the Gates-Learjet Company) retain the stamp of the original Learjet's look.

Bill Lear—big, often profane, sometimes maddening to work with, tremendously loyal to his employees—was not dismayed by criticism when he announced what proved to be his final product, the Lear Fan. This was to be as great an advance as the Learjet had been, not in speed but in configuration and structure. Lear's goal was to build and certify an aircraft distinguished by utter cleanness of line, manufactured from advanced composite materials, and employing two turbo-

**The Israel Aircraft Industry LAVI is one of the most advanced and controversial fighters of this decade and a good illustration of a smaller country's ability to create fully competitive aircraft.**

**OVERLEAF:**
**The three-engine Dassault-Breguet Falcon Jet is built to the same standards of quality and performance as the Mirage fighters.**

prop engines to drive a single pusher propeller. Lear saw a niche in the market that could be filled by relatively low-cost, economical aircraft capable of providing high speeds and luxury.

The Lear Fan went through an agonizing series of crises. It flew well, but problems with the transmission system (shades of J. V. Martin's Cruising Bomber) and with the structure itself delayed certification. Bill Lear died of cancer while the program, despite its problems, still seemed to have a chance. His wife Moya valiantly carried on and several times seemed to be on the verge of achieving the certification that would have allowed production to begin. Unfortunately, the Lear Fan project ran out of money before it ran out of problems, although it may yet be revived.

Despite its failure, the Lear Fan has provided the inspiration for a number of other radical shapes and radical structures, some of which now represent the leading edge of the art. It deserves a place in history just below the Learjet itself because it pointed the way to where we are today.

# 9. TODAY AND TOMORROW

Successful flight began with two brothers of genius concentrating on the fundamental areas of configuration, materials, structure, powerplant, and controls. Their final choice in each was perfect: a biplane construction for flexible strength, with a canard surface to soften the stall; wood, fabric, and metal materials, each correct for its function; an engine generating barely sufficient power through a transmission to propellers of adequate efficiency; and three-axis control to overcome the birdlike instability that they had the piloting prowess to master.

Today and tomorrow, in all regimes, the leading edge of flight will be found focused on these same five areas. Vast quantitative differences separate present and future efforts from those of the Wrights—speeds are hypersonic, altitudes are stratospheric, power is abundant, and controls are computerized—but the Wrights were first, and their flight defined all aviation.

Existing and prospective leading edges embrace a wide array of disciplines, including sport flying, general aviation, executive aircraft, commercial air transport, and military aircraft.

## SPORT FLYING

After the unexpected ease with which they solved the primary question of flight, the Wrights immedi-

**RIGHT:**
Hang gliding provides entry into flight for many otherwise excluded.

ately turned to the military services in order to broaden the market for their invention. The airplane became primarily an instrument of war for many years, and not until the mid-1930s did it become profitable as a commercial vehicle. Sport flying was always left to the relatively few rich individuals who could afford to buy and maintain an airplane.

Today, a loyal band of several hundred thousand pilots are desperately trying to put the sport back into flying. They labor against formidable financial and regulatory obstacles, but they are succeeding beyond all imagination and (perhaps

**Today's ultralights, like this Eipper, offer another way into aviation.**

more than any other factor in aviation) offer hope for the future.

No other sport possesses a counterpart to the grand array of ultralight aircraft now buzzing like swarms of dragonflies around small-sized airfields across the country—a massive grass-roots rebellion against rising costs, increased regulations, and excessive sophistication. In a way, the ultralight proponents are turning technology against itself, using advanced materials and ideas to offset the high prices that are bogging down the rest of aviation. The ultralights were originally no more than hang gliders to which a chainsaw engine and club propeller had been strapped; and hang gliders themselves were a heroic step backward in time to the days of Otto Lilienthal.

The transition from hang glider to ultralight was accomplished in a climate of idealistic optimism, pervaded by a sense that a corner of flying had been overlooked and was just now being discovered—an arena where training, licensing, and regulation would not be necessary. Because the movement initially involved relatively few people, the Federal Aviation Administration was content to ignore the field and not call for the sort of regulation that encumbers—necessarily, perhaps—all other aspects of aviation.

As a result, in the early days, ultralight aviation was exploited by everyone. Would-be pilots exploited the freedom, and manufacturers exploited the would-be pilots. A wild series of designs appeared, some of which were sold as kits even before the prototypes flew. Aggressive, assertive people—the worst kind for the purpose—felt that they could build a kit and teach themselves to fly, insulated by the illusory protection of an alluring freedom.

There were tragedies, of course. Poorly designed ultralights came apart in the air. Others, suffering from a hard landing or from being improperly assembled, broke up. In some cases, ill-trained pilots took well-designed ultralights past their limits and crashed. A well-known television newsman did exactly that, on camera, to the great detri-

Tom Poberezny is not only one of the world's leading aerobatic pilots, he is a capable manager following in his father's footsteps.
**BELOW LEFT:**
Paul Poberezny is the driving force behind the resurgent home-built aircraft movement.

ment of the movement. The beginning motorcyclist syndrome asserted itself; when new ultralight pilots acquired a little proficiency, they would replicate all the stupid things done by pilots in the past—buzzing, low-level maneuvers, aerobatics—again with predictable results.

In a relatively short period of time (but a painfully long one, in terms of casualties), the ultralight movement began to put itself in order. The primary point that had been denied in the initial aura of good will was that the ultralight was in fact an airplane, and not some Daedalian miracle of wings that granted initiates both the power of flight and the ability to use it. When this was recognized, ethical manufacturers declined to sell kits to buyers who had not received proper training. At flying fields, a sense of discipline emerged, in which it was *de trop* to be careless, not to inspect the aircraft, or to attempt to exceed the limits of the person's flying skill. Concurrently, the FAA moved inexorably toward some level of regulation.

Now the tide is turning, and the movement stands to benefit from new materials that have filtered down from more advanced fields of aviation. An entirely new type of aircraft has emerged: the ARV (Aircraft Recreational Vehicle), which establishes a needed intermediate step between the ultralight and the light plane. Smaller, lightweight versions of such established light plane designs as the famous Cubs and sleek new models tailored to the composite materials only recently available are now on the market.

Still, the specter of litigation hangs over ultralights, as it does the light plane industry, and some manufacturers have been forced to declare bankruptcy after accidents occurred. It may be that future sales will have to be made with a hold-harmless agreement—not as part of the sales contract, but as part of the vehicle title itself—if the industry is to survive.

Potential liability even haunts what had been the most independent and promising flying constituency in America. Thousands of home-built aircraft are flying, and tens of thousands are being built; but now many kit producers and even

blueprint designers are folding their tents, unwilling to risk the threat of a multimillion-dollar suit for damages in the event of a crash.

And yet much remains positive about the home-built aircraft movement, from the wild variety of designs to the strength of the participation to the extraordinary craftsmanship designated as standard by the Experimental Aircraft Association.

The EAA began twenty-six years ago in Paul Poberezny's basement; the genial and far-sighted founder has since seen the association blossom to include over 100,000 members, in all parts of the world. Under Poberezny's guidance, the home-built aircraft movement has expanded to include ultralights, classic aircraft, antique planes, and warbirds (the last class ranging from an L-5

**No man has captured the purity of flight as well as Paul MacCready, whose Gossamer Condor led to a series of aircraft including this solar-powered Gossamer Penguin.**

Cub to flying B-29s and a variety of jet fighters).

Poberezny's success can be attributed to many things, but underlying all else is an absolute requirement for excellence in everything the association has attempted, from building airplanes to cleaning up after the hundreds of thousands of visitors who attend the fly-in held at the EAA's home base at Wittman Field in Oshkosh, Wisconsin. Each year, the crowds grow in numbers and enthusiasm. In a single week, as many as 10,000 airplanes, from ultralight to Concorde, pour into Wittman Field, completely saturating the parking areas and the entire city of Oshkosh. The miracle lies not only in the craftsmanship and variety of the airplanes (of which more later), but in the incredible good humor and courteous behavior of the crowds.

The annual fly-in begins with an electrifying opening of crack aerobatic teams, smoke-dispensing parachutists, and rousing national anthem. It then proceeds through intricately marshaled flybys—a choreographed aerial ballet of continual, simultaneous takeoffs and landings of World War I fighters, World War II bombers, resplendent home-builts, supersonic transports, and pedal-powered blimps, all in a good-natured show-and-tell procession that serves as the raison d'être for many participants. The intense, agonizing judging for the championship of each of the dozen classes of entrants then ensues. It is unquestionably the greatest air show on earth.

Curiously, despite the thousands of aircraft—those buzzing, noisy, attractive mechanical devices—the real joy is in the human interaction. The place is alive with people anxious to teach and people anxious to learn. This endless swapping of information is perhaps the most egalitarian event of its kind. Governors mix with brain surgeons and filling-station attendants; and a filling-station attendant who has special knowledge of welding, painting, or laying up fiberglass becomes the guru for the others.

The airplanes vary in size, style, and construction, encompassing everything from lovingly restored 1932 Waco cabin planes to gossamer-winged ultralights to massive formations of North American T-6s or F-51s. All-wooden airplanes, all-metal airplanes, and all-composite airplanes are in evidence, as are planes of every conceivable mixture. The only common denominator is a loving attention to detail and finish. Each airplane is a jewel, burnished and nervously attended to by its owner at all times.

The EAA has been augmented by a new museum, which was transformed from a cornfield to the finest sport aviation museum in the world in a single year of brilliant bustling activity. But the association depends foremost on the strength and support of its active membership, which brings an enthusiasm to the movement that would be impossible to purchase. It also brings a conviction that America still wants to fly and is willing to spend the money and the thousands of hours necessary to create an airplane in which to do it. Among the many fine things the EAA represents is a clear signal to the American aircraft industry that a market exists for general aviation airplanes, if the necessary courage and the innovation can be summoned to develop it.

## General Aviation

It is a bitter thing to say, but general aviation—the steady production of better and better aircraft from famous factories in Wichita and elsewhere—is moribund. More than 700,000 registered pilots live in the United States, and more than 210,000 registered aircraft are stationed at 14,000 public and private airports. Yet general aviation is in desperate straits; the production lines have halted for all but expensive executive-type aircraft, prices have skyrocketed, and the industry shows a pervasive lack of will to deal with the problem.

This is not a parochial problem. In 1979, 17,032 aircraft were sold by the major American light plane manufacturers. These aircraft generated billions of dollars of profits and were the source of hundreds of thousands of jobs. The benefits from aircraft production do not end at the Wichita factory; they extend in vast, widening ripples across the United States. In 1984, only 1,995 aircraft were produced. It was as if Detroit had slipped not from 10,000,000 cars per year to 7,000,000, but all the way to 1,200,000.

The situations of Wichita and Detroit significantly differ only in the matter of quality. While the Detroit iron was allowed to become ludicrously large and was subsequently overpowered with a proportionate loss in quality control, Wichita became ossified but somehow maintained its high quality of production. Unfortunately (and unlike the situation with airline transports), the aircraft being made had not changed except in detail—primarily in paint schemes—in twenty years. Reasons have been given for this, but the principal cause was concern over liability. If something worked, Wichita was determined not to fix it,

even if it meant staying with the same materials, airfoils, and engines for decades.

This situation contrasts markedly from the automotive illustration: Detroit at least did not standardize and keep building 1960 Chevrolets and Fords. But notwithstanding superficial changes, this is exactly what happened in Wichita, where basic designs such as the Cessna 172, the Beech Bonanza, and the Piper Comanche became permanent parts of the aviation scene. At the same time, prices went up and up, far beyond the inflation rate.

There are creaking signs that things are changing—signs introduced at the only level at which the industry is continuing to function, that of the

The ultralight movement is replicating the development of aircraft. Planes like this CGS Hawk are very similar to "real" airplanes in feel and comfort.

executive transport. Even though sales are down overall on the aircraft used for transporting executives, the manufacturers have preferred to concentrate on this market to the exclusion of the private pilot. The underlying considerations are twofold: return on investment and exposure to liability. The profit to be made on a $22,000,000 jet is obviously higher than that to be made on a $100,000 private plane, and the chance of an accident is far less. Executive transports (which can be anything from a single-engine Beech to a Gulfstream IV) are typically flown by well-paid professionals who maintain a high degree of proficiency in flying the aircraft, receive frequent simulator training, and pass semiannual flight checks. The airplanes themselves generally receive more systematic and thorough maintenance than might be given to a light plane. The chance that a legal problem will arise is thus vastly reduced.

It is encouraging for the industry as a whole that this somewhat exclusive class of aircraft is also the focus for revolutionary change, both from new suppliers outside the industry and from an industry stalwart, Beech Aircraft. The new designs embrace all five of the fundamental areas explored by the Wrights. Configurations are changing radically, with canard surfaces and pusher engines appearing on a number of designs. The use of composite materials is increasing, and the structures are far more advanced. Some even have the appearance (if not the reality) of being biplanes. Finally, the new breed is pushing into regimes of altitude and speed that had previously been primarily the province of military types, commercial air transports, or the most expensive executive jets.

The use of composite materials that are laid up and treated with certain other materials and then allowed to cure, either through their own chemical reaction or by baking in some sort of autoclave, goes back through all of aviation history. A whole series of processes come to mind, ranging from the Haskelite methods developed after World War I to those used by John Northrop in the Lockheed Vega to the aircraft materials and processes introduced by Howell Miller, Otto Timm, Vance Breese, and of course Virginius Clark (with his Clark F-46 and Hughes HK-1 triumphs). The current crop of materials encompasses the familiar fiberglass to very advanced materials used as components on modern fighters.

Dr. Leo Windecker created and brought the Windecker Eagle to full FAA certification in the 1960s, backed heavily by the Dow Chemical Company. Bill Lear's Lear Fan, however, deserves credit for starting the current trend toward composites in a whole series of innovative prototypes.

In lovely Camarillo, California, another group of experienced high-tech professionals brought forth a wildly beautiful design, the AVTEK 400. Perhaps

Sophisticated twin-engine aircraft like this Piper Seneca remain the heart of business aviation efforts.

**OPPOSITE:**
In businesses, the accountants scrub the investment in a business aircraft like this Learjet just as they do investments in machine tools or buildings. The aircraft must be shown to produce profit or it goes.
**OVERLEAF:**
The Voyager project embodies the best of the American dream of flight. Three dedicated individuals—designer Burt Rutan and pilots Dick Rutan and Jeana Yeager—assembled a small team of volunteers and established a record nonstop, non-refueled flight around the world. The aircraft is as radical in structure as in configuration and signals a new era in aviation.

the most impressive of the AVTEK's credentials are its principals. They include the amiable, farsighted Dr. Leo Windecker; the versatile pilot-designer-engineer-dreamer Dr. Paul MacCready, whose work on the Gossamer series of human-powered aircraft brought low-speed aerodynamics into the prominence it deserves; and the legendary veteran Al Mooney, of the Mooney Aircraft Corporation. Backed by Dow, Du Pont, and a group of seasoned investors headed by Robert Adickes, the AVTEK has a substantial chance to break into the marketplace.

The sleek airplane is built primarily of Kevlar and Nomex honeycomb, two lightweight, high-strength aramid fibers developed by Du Pont. Carbon-fiber composite material is used in the AVTEK's spars and in other places where maximum strength is required.

The AVTEK has a forward canard mounted above the cabin in biplane position and two pusher propellers. It boasts an empty weight of 3,300 pounds, about half that of conventional twin-engine turboprops. It is built to compete directly with the Lear Fan and reputedly is capable of generally higher performance.

The most advanced executive aircraft on the horizon, however, is one that combines a radical configuration with full use of composite materials; its glamour is enhanced further by its marriage of the talented and innovative designer of home-built kits, Burt Rutan, with the Rolls Royce–style conservatism of Beech Aircraft.

The Beech tradition of excellence goes back to the engineering skills and piloting genius of the company's founder, Walter Beech. His wife, Olive Ann, one of the first of the really powerful women executives, brought stabilizing business acumen to the firm that Walter founded in 1932. By creating the best aircraft of the time for its market—the still-coveted Beech Staggerwing and the twin-engine D-18—Beech emerged as the premier manufacturer of business aircraft. The company expanded its role during World War II to many other areas, but then returned to solidify its position in the marketplace with a long series of classic designs

that began with the V-tail Bonanza of 1946.

*Beech* came to mean quality, solidity, and evolution rather than revolution. Yet a time came for change, when management had to look beyond the incredible series of past successes into the future of diminishing oil supplies and increased competition from abroad. Beech's management turned to a young genius whose career took off in the field of home-builts but who had defined his engineering methods to a point where he was sought after by major manufacturers.

Burt Rutan is a professional engineer, seasoned by working with the U.S. Air Force and NASA and by a stint with Jim Bede, one of the most controversial of all aviation personalities. He established the Rutan Aircraft Factory (RAF) at the Mojave Airport in California and commenced a series of radical designs whose use of the canard configuration and composite materials stood the home-built movement on its ear. The first of Rutan's designs was the Vari-Viggen, in 1968, an aircraft somewhat inspired by the Swedish Viggen canard fighter but employing conventional mixed-construction techniques.

Thereafter, a series of increasingly advanced designs poured like popcorn from the tiny factory. The VariEze reestablished the canard configuration in the world's mind and used all-composite glass-foam sandwich construction. Its dragonfly appearance and startling performance provoked a flood of orders. Lessons from the field permitted an improved version, the Long-EZ, which had an even bigger market and jet-fighter flight qualities. The Defiant was next, a twin-engine, executive-type transport that went from sketch to flight in seven and one-half months.

Other designs—some for kits, some for production, some for experimentation—tumbled from the factory, including the unaesthetic but powerful Grizzly, the tiny 18-horsepower Quickie, the Solitaire motor glider, the Amsoil racer, and (most ambitious) the Voyager.

In December 1986, Burt Rutan stunned the public with the unbelievable nonstop, non-refueled flight around the world of his Voyager. The radical Voyager was clearly a Rutan design with its all-

**OPPOSITE:**
An intermediate step to the Beech Starship was the "push-pull" Rutan Defiant, a twin-engine executive aircraft for the homebuilder. It links the Starship to the VariEze and other earlier Rutan products.

composite construction, twin booms, canard surfaces, and ultra light weight. Built, tested, and flown by Dick Rutan, his brother, and Jeana Yeager, the Voyager's dramatic nine-day flight thrust aviation into a future full of new ideals and challenges.

Although more of Rutan's designs are now flying worldwide than any other brand of home-built aircraft, they represent only a part of his productivity. He also established a firm called Scaled Composite Via Advance Link to Efficiency Development—SCALED COMPOSITE—that developed scaled-down versions of proposed full-size aircraft at unbelievably low prices. He built an oblique-wing prototype for NASA for $250,000; it subsequently emerged that the bid had been approved routinely and without question by NASA procurement in the belief that it was for a scale model. Had it been known that the amount was for a piloted test vehicle, procurement would probably have rejected it as an "unrealistically low bid." In a similar way, he created the 62-percent-scale Fairchild NGT (for Next Generation Trainer), which helped Fairchild win the Air Force T-46 competition.

Early in the 1980s, the then president and chief executive officer of Beech, Linden Blue, went to Rutan in strictest confidence and asked him to develop the executive plane of the 1990s—and the twenty-first century. Rutan explored dozens of configurations, some conservative, others even more radical than usual, before delivering (again in total secrecy) an 85-percent-scale version of the Beech Starship. Beech rocked the National Business Aircraft Association convention in 1983 with not only a filmed presentation of the new aircraft, but a flyby, takeoff, and landing demonstration of the beautiful, radical, high-performance Starship.

It was as if Rolls Royce had suddenly announced a gas turbine car, or Tiffany a new kind of diamond. Beech has committed itself to a billion-dollar effort to build the Starship, and Rutan has since moved into an executive position with the company.

The Starship has all of the Rutan (and some of the Wright) trademarks: twin pusher turbo-props mounted on a leaned-out delta rear wing; wing-

lets (tipsails at Beech); and an innovative, variable-sweep canard, called a forward wing.

The result is a high-performance (better than 350-mph cruise speed), stall-free, quiet executive aircraft made of carbon-fiber skins applied to a structural core of Nomex honeycomb panels. Titanium will be used in high-stress areas, and an aluminum mesh will be imbedded in the composite skin as protection against lightning. The fuselage will be mass-produced on a giant filament-wound rig capable of making a complete fuselage in twenty-eight hours (compared to the nine days required with conventional methods).

Beech's massive commitment to a new technology will yield an entire family of designs. And sig-

**The Starship combines a host of new features from all-composite construction to the canard configuration. It is the single biggest development program ever undertaken in business aviation.**

nificantly, the entire step forward had its basis in the home-built aircraft field. The effort will also produce spillovers to the commercial airline market, which is girding itself for a launch into the twenty-first century.

## COMMERCIAL AIR TRANSPORT

The Boeing 767 represents the current state-of-the-art in economical commercial transport. But the world's commercial air transportation manufacturers—Boeing, McDonnell Douglas, Airbus Industrie, and Lockheed—are on the cusp of an economic paradox. The wild increase in oil prices during the 1970s prompted the builders to turn to

The cockpit of the Starship looks like it could carry it to at least "warp five."

fuel economy in their next generation of airliners. Gasoline prices were expected to worsen so that 1978's $1.50 per gallon would be $2.50 per gallon or more by 1985. The resulting series of design projections have been waylaid by the decline of OPEC solidarity and the continuing dip in fuel prices.

Oddly enough, this drop in prices has made existing aircraft already in production much more attractive than before. Boeing 737s and McDonnell Douglas MD 80s, when fitted with new engines and flight decks, can achieve almost all the advantages intended for the next generation of commercial transport aircraft without the billions of dollars of capital expenditure.

The issue has not been fully resolved as yet.

American manufacturers are saying that the investment should not be made merely for the sake of change. On the other hand, Airbus Industrie is pressing ahead, seeing a market that the American firms are choosing to ignore.

Advances in engine technology make the decision over whether to embark on an advanced design now or later even more difficult. The primary focus is on an engine type that more nearly resembles a throwback than an advance, because of its large, multiple-bladed, curved fan. But the concept is a logical step forward from the high-bypass-ratio fanjets of the 1970s. In its several variations, it is called a prop-fan, a free fan, or an unducted fan, depending upon the manufacturer's way of connecting the fan to the engine. In some (Rolls Royce and Pratt & Whitney), the fan has a gear system; in the General Electric engine, a small gas generator provides power straight to the turbines, which are connected directly to the fan. All aim at the same thing: mass movement of air for thrust.

The new engine, whose outline resembles that of a small, rubber-bladed tabletop electric fan from the days before air conditioning, promises savings of 25 percent in fuel consumption and a power growth potential to as much as 75,000 pounds of thrust.

If, as expected, engineers are able to overcome problems of vibration, change in relative angle, flutter, and noise in the unducted powerplant, a Rutanesque airliner of 150 to 200 seats may emerge in the mid-1990s. A large canard surface forward and relatively small swept wings (equipped with winglets) aft would be mounted on a conventional-looking fuselage. The twin engines would be placed like those on a contemporary DC-9, but far enough out from the fuselage for the paddle-bladed propeller to have clearance.

The resulting aircraft would have as much as a 60 percent improvement in fuel consumption, deriving from the engine, from the improved wing design (possibly with a laminar flow control), and from lighter weight. The structure would contain substantial amounts of aluminum/lithium material and advanced composites for weight reduction.

Further economy would result from the reduction in aircraft control surface size permitted by active controls—already in service in the Lockheed 1011s and regarded as one of the most promising areas for future developments. In an active control system, a computer linked to the autopilot automatically supplies control inputs that provide gust alleviation, stress reduction, and improved fuel consumption. Future systems will permit stability to be relaxed and compensated for by a longitudinal stability system. This will allow engineers to reduce drastically the size of the horizontal control surfaces, with a consequent reduction in both drag and weight.

A passenger would only be aware of the

RIGHT:
The Boeing 747F. Freighter aircraft have not yet come into their own, but the next twenty years will see a change.

changes indirectly, from a stabilization (and perhaps even a reduction) in fares and by sensing that the ride characteristics had been improved when turbulence was encountered.

The more distant future holds dramatic changes, including aircraft of significantly different shapes and sizes, and revolutions in ground support systems, approach controls, and every other aspect of flight. The primary driver for the new breed will be economics; instead of concern about reaching Europe from the United States, or vice versa, the concentration will be on connecting those two places with the great Pacific basin. In effect, the aim will be to match the current three-hour Concorde time between the New World and the Old with three-hour flights from London to Sydney, from New York to Beijing, or from Bonn to Singapore.

To do this will require not SSTs but HSTs: hyper-

sonic transports riding their own shock wave and burning not petroleum-based fuel but either liquid hydrogen or liquid methane. They may even have a new acronym—TAV, for transatmospheric vehicle.

The new transports, which are already being called "the Orient Express," will be much like passenger ballistic missiles, wiping away barriers of time and distance in a wedge-shaped vehicle with enormously powerful hybrid engines that adapt themselves automatically from hypersonic (as much as Mach 12!) speed to slower, conventional speeds for approach and landing. The airplane will be linked to ground computer networks that convert the crews into airborne supernumeraries, standing by for emergencies only.

The path to this mid-twenty-first-century technology will be through the development of the next two generations of military aircraft.

## MILITARY AIRCRAFT

Not surprisingly, every element of progress heretofore mentioned is either already flying or on the drawing boards for military fighters and bombers. The precursor of the next generations is perhaps the Grumman X-29, an airplane remarkable for its configuration, its control, and its combination of components.

The aft-swept wing has been a familiar scene in the sky since it was introduced on the Messerschmitt Me 163 and 262 fighters. The swept wing had some stability and control disadvantages at lower speeds, however, including pitch-up, loss of lateral control, and roll-off at the stall. From the beginning, it was understood that a forward-swept wing could avoid these problems, and this was demonstrated by the Junkers Ju 287, a very advanced swept-wing bomber flown successfully in 1944. Forward sweep permits the wing's carry-through structure to be placed aft of either the bomb bay or the passenger compartment, and in the case of the bomber, minimizes trim changes on weapon release.

But disadvantages with forward sweep become evident at high speeds. The forward-swept wing of the Junkers Ju 287 tended to twist at the tip, causing an upload that could be catastrophic if the structure were not heavily reinforced. In a metal wing, the extra weight required for reinforcing offsets the advantages of forward sweep, and we have had forty years of aft-swept wings as a result.

With materials now available, however, engineers could create aeroelastically tailored composites to overcome the tendency of forward-swept wings to twist, at very light weights. A less obvious, but militarily far more important gain of the forward-swept wing is the higher lift coefficients it offers in maneuvering flight—that is, in dogfighting.

The Grumman X-29 is intended to be a technology demonstrator; for economy, major components of existing aircraft were adapted. The Northrop F-5A front fuselage and nose gear was combined with an off-the-shelf General Electric F404 engine and a main undercarriage from a

General Dynamics F-16. The X-29, in addition to its forward sweep, uses a wing of thin, super-critical section, made of graphite-epoxy. A metal wing of the same thickness would twist like a hack-saw blade at high speed. A variable-incidence, close-coupled canard surface is used to minimize trim drag.

To take full advantage of the design possibilities, the X-29 was designed for neutral longitudinal stability in supersonic flight, meaning that the pilot must depend upon a three-channel, digital, "fly-by-wire" system that senses deviations from desired flight conditions and applies instantaneous corrections far more quickly and sensitively than the pilot can. If by some chance all of the computer systems should fail at a speed of over 200 knots, the X-29 would pitch up and down far faster than the pilot could react, and within two-tenths of a second it would completely disintegrate.

As advanced as it is, the X-29 is but a prelude to the future. Manufacturers all over the world are adapting conclusions drawn from the X-29 program to new fighters.

The pressure is caused by the parity recently established by the Soviet Union's Sukhoi Su-27 Flanker and MiG 29 Fulcrum, which are very close in quality to the United States Air Force's McDonnell Douglas F-15 and General Dynamics F-16. For the past decade, the United States has counted on the qualitative superiority of its fighters to off-set the numerical superiority of the Soviet Union's air fleet. Future parity depends on the Advanced Technology Fighter.

A multi-billion dollar fighter program for the twenty-first century is coming to a boil with the roll out of the Lockheed YF-22A and Northrop YF-23A advanced tactical fighter prototypes. These sleek new jets promise enormous advances in speed, range, and maneuverability and will also incorporate the latest concepts in stealth technology.

Such advanced aircraft are so expensive that airframe manufacturers must form consortiums to be able to contend for contracts. Lockheed is teamed with Boeing and General Dynamics on the YF-22A, while Northrop is allied with McDonnell

The next century may see a National Aerospace Plane, looking something like this artist's conception, lead the way for a family of trans-atmospheric vehicles.
**RIGHT:**
The Grumman X-29—like almost all combat aircraft of the future—is totally dependent on its computers to achieve stability.

Douglas for the YF-23A. The two major jet engine companies are also competing for this critically important contract. Both aircraft will be tested with Pratt & Whitney YF119 and General Electric YF120 engines.

The success of the new fighter, no matter who builds it, will hinge on a massive national effort to create a host of new systems including avionics with very-high-speed integrated circuit technology, new missiles, and advanced engines of greater power, possibly using vectored thrust, yet having a lower specific fuel consumption and decreased infrared signature.

The functioning of the entire aircraft will of course depend upon advanced adaptive aerodynamics; fittingly, the Mission Adaptive Wing being developed by Boeing under a combined NASA/USAF program returns to what might simplistically be called wing-warping—the very point where the Wrights started!

The Wright brothers drew many inferences from the shape and structure of wings. The control expressed in the darting flight of hawks and eagles is effected by their brain control of the myriad surfaces. The first attempts to follow this example was wing-warping, replaced subsequently (for weight and structural reasons) by ailerons.

Much later in the progress came a variety of leading and trailing edge flaps to improve lift/drag characteristics, followed by the variable-sweep wings found on the Grumman F-14 Tomcat and General Dynamics F-111 fighter.

The Mission Adaptive Wing (MAW) is designed to achieve all of the goals of the intermediate steps by changing shape in flight to maintain optimum aerodynamic performance at all times. The MAW has no hinged flaps, spoilers, or projecting surfaces to break the smooth contour of its upper surface. Advanced variable-camber mechanisms, coupled with digital flight control computers and sensors, regulate the contour of the fiberglass material that covers the wing. Camber (the airfoil shape) can be varied during flight from leading edge to trailing edge and from wing tip to wing tip —a sort of warping squared!

Initial tests show that the MAW will improve buffet-free lift by 69 percent and sustained lift by 25 percent, while reducing supersonic and subsonic drag by 7 and 6 percent, respectively. The wing will automatically trim its profile to increase range by as much as 30 percent in the cruise condition. In dogfighting, the MAW will permit very high G turns to be made without exceeding structural limits. Instantaneous turn rate, the crucial factor in dogfighting, will be increased by an estimated 20 percent. The MAW will be more expensive to construct, but its longer structural life (due to automatic load alleviation) will result in lower life-cycle costs.

The fighters that proceed from the Mission Adaptive Wing study will be with us through the first half of the twenty-first century, at least, and perhaps longer. Part of their duty will be to protect bombers and transports of almost as exotic configuration.

The first of these, the Northrop B-2 stealth bomber, has instantly become one of the most controversial bombers in history. A flying wing designed to be virtually invisible to radar, the expensive B-2 may yet prove to be the weapon of choice in a world in transition from a dying Cold War into a period of unending crisis in the Middle East.

In the next century, we may see the X-31, the National Aerospace Plane (NASP) lead the way for a family of trans-atmospheric hypersonic vehicles, some with orbital capability.

Far below these incredibly fast superplanes, we may have lightweight sport planes that are capable of flying as slow as birds, are crashworthy, are relatively inexpensive—and perhaps are of a nature to lure multitudes back into aviation.

The Panavia Tornado is another product of European cooperation (*top*). Modern cockpits increasingly take on a *Star Wars* appearance (*bottom*).

## THE HUMAN SPIRIT: ALWAYS THE TRUE LEADING EDGE

No matter what the aircraft of the future look like, no matter how expensive they are, no matter how they affect our lives, they will have a basic and vital connection with all other leading edges that have appeared over time: each will be founded upon a scientific advance whose ultimate source was a human mind. Other familiar patterns will also remain in force. The Advanced Technology Fighter of the United States will have its European and Soviet counterparts, each with a similar performance, a similar history, and a similar human grounding. Simultaneous successes and simultaneous failures will occur; alternative and perhaps better solutions will be abandoned through politics; opportunities will be lost and avenues closed. No matter what technology, no matter what country, the role of the human spirit in leading the way has international application.

Leading edges have always appeared over time as the result of the needs of the moment and the inclination of engineers and scientists to be discontent with current accomplishment. In retrospect, the geniuses of the past can be recognized and saluted, operating as they were within the framework of their own national societies; but in the future, the leading edges will be beyond the scope of individual nations, and to gain the edge nations will be required for economic and technical reasons to join together in the development.

The Concorde of the 1960s, produced through the harmonious working relationship of England and France, could well be matched by the hypersonic transport of 2010, built by a Japanese/Chinese/American/European/Russian consortium. From there, it would be only a step to a jointly operated space station, and thereafter perhaps world peace and permanent security. Interestingly, that was what the Wright brothers thought their leading edge, the Flyer, might bring about. The leading edge of the twenty-first century may be one of global consensus; and if it is, it will depend as always on the human spirit.

222

**LEFT:**
The saw-tooth shape of the Northrop B-2 stealth bomber combines with exotic structural materials to foil enemy radars. The B-2's many technological advances presage a brilliant era of flight in the twenty-first century.

**OPPOSITE:**
Perhaps the best-kept military secret of all time, the Lockheed F-117A stealth fighter is equally potent as a political or military weapon.

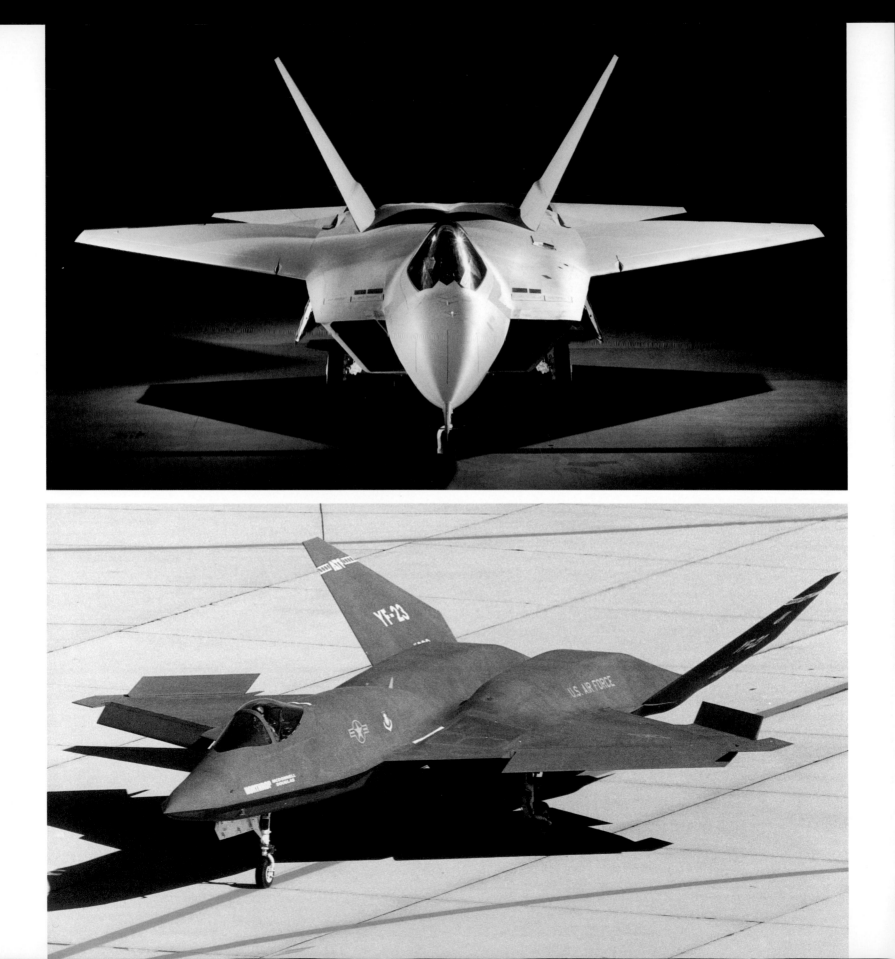

**OPPOSITE:**
The Lockheed YF-22A advanced tactical fighter is a joint product of a Lockheed-Boeing-General Dynamics consortium. A production run of 750 aircraft will yield a unit cost of about $51 million (*top*). A Northrop-McDonnell Douglas team has fielded the competitive YF-23A, whose shape reflects lessons learned from the B-2 (*bottom*).

**RIGHT:**
The McDonnell Douglas/General Dynamics A-12 is a carrier-based, long-range penetration bomber. The pure flying-wing shape of the "Avenger" came as a surprise to the industry.

**BELOW:**
The Boeing/Bell V-22 Osprey is a valuable military weapon—and an invaluable solution to the city-center-to-city-center civil transport problem.

**BELOW:**
Western observers have been astounded by the superlative performance of the Sukhoi Su-27, the Russian equivalent of the F-15.

**RIGHT:**
About equal in size to the McDonnell Douglas F/A-18 Hornet, the MiG 29 has been demonstrated at air shows around the world in a reversal of past Soviet policy.

# INDEX

# PHOTO CREDITS

**Page 1:** © Erik Simonsen
2: courtesy Art Davis
4: courtesy Terry Gwynn-Jones
7: © 1982 Dick Durrance/Woodfin Camp & Assoc.
8–9: courtesy Terry Gwynn-Jones
10: National Air and Space Museum
11: © Aram Gesar/Gesar Inc.
12–13: © Herman J. Kokojan/Black Star
14: National Air and Space Museum
17: © 1981 James Sugar/Black Star
18 (both), 19, 20–21: National Air and Space Museum
22: © Frederic Lewis/Harold M. Lambert
23 (all), 24: National Air and Space Museum
25: The Bettmann Archive
26–27: Michael E. Long
28: © 1982 Russell Munson
29: courtesy Art Davis
30: © Erik Simonsen
31: Michael E. Long
32: © 1981 Russell Munson
32–33: courtesy Art Davis
35: Culver Pictures
36, 37, 38, 39, 40–41: courtesy Terry Gwynn-Jones
42: National Air and Space Museum
43: courtesy Terry Gwynn-Jones
44 (top): courtesy Messerschmitt-Bölkow-Blohm GmbH
44 (bottom): National Air and Space Museum
45 (top left): courtesy Messerschmitt-Bölkow-Blohm GmbH
45 (bottom left, top right): © Howard Levy
46: courtesy Messerschmitt-Bölkow-Blohm GmbH
47 (both): courtesy Terry Gwynn-Jones
49: © Cliff Feulner/The Image Bank
50–51: © 1978 Gary Gladstone/The Image Bank
52, 53: © Frederic Lewis
55: © 1978 Gary Gladstone/The Image Bank
56, 57: The Bettmann Archive
58: © 1979 Cliff Feulner/The Image Bank
59: © 1978 Gary Gladstone/The Image Bank
60 (all): National Air and Space Museum
61 (top): © Frederic Lewis
61 (center, bottom), 62–63: © Howard Levy
64–65: © 1984 Frank Whitney/The Image Bank
66 (both): © Frederic Lewis
67 (all): Howard Levy Collection
68 (both), 68–69: courtesy Musée de l'Air et de l'Espace
70–71: © David Burnett/Woodfin Camp & Assoc.
71: courtesy Air Force Magazine
72–73: U.S. Navy Photo
74, 75 (top): National Air and Space Museum
75 (bottom): The Bettmann Archive
76–77: NASA
78 (top): courtesy Terry Gwynn-Jones
78 (bottom): © Frederic Lewis/Edwin Levick
79: Kirby Harrison
80–81: © 1985 Russell Munson
82 (top): National Air and Space Museum
82 (bottom), 83 (top): courtesy Terry Gwynn-Jones
83 (bottom): The Bettmann Archive
84: © 1985 Russell Munson
**Page 84–85:** © Cliff Feulner/The Image Bank
86–87: © 1984 Russell Munson
88–89: National Air and Space Museum

90–91: © Dan Moore
92: courtesy R. S. Allen Collection
93 (top): courtesy Grumman Corporation
93 (center): courtesy R. S. Allen Collection
93 (bottom): courtesy Grumman Corporation
94: R. Arnold Collection/National Air and Space Museum
95: © Lionel Isy-Schwart/The Image Bank
96–97: National Air and Space Museum
98–99: © 1985 Russell Munson
100: courtesy Underwood Collection
101: courtesy R. S. Allen Collection
102–103: courtesy Jay Miller/Aerofax, Inc.
104–105: © 1984 Russell Munson
106 (all): National Air and Space Museum
107: © Howard Levy
108, 109: National Air and Space Museum
111: © Aram Gesar/Gesar Inc.
112–113: courtesy The Shuttleworth Collection
114–115: © Douglas Kirkland/The Image Bank
116 (bottom left): courtesy Underwood Collection
116 (top right): The Bettmann Archive
116 (center right), 117: courtesy Underwood Collection
118 (top): courtesy EAA Aviation Center/Ted J. Kostan
118 (second from top, third from top, bottom): © Howard Levy
119: © 1985 Russell Munson
120 (both), 121 (both), 122 (all), 123, 124 (left): National Air and Space Museum
124 (right): © Robert DeGroat
125: R. Arnold Collection/National Air and Space Museum
127: National Air and Space Museum
128–129: © Dan Moore
130–131: © 1985 Mike Fizer
132 (both): National Air and Space Museum
132–133: © Aram Gesar/Gesar Inc.
134: © 1981 George Hall/Woodfin Camp & Assoc.
135 (top): courtesy Lockheed Corporation
135 (center): courtesy The Boeing Company
135 (bottom): National Air and Space Museum
136–137: © 1981 James Sugar/Black Star
138: courtesy Frederick A. Johnsen
139: © Herman J. Kokojan/Black Star
140–141 (all): courtesy Jay Miller/Aerofax, Inc.
142: © 1985 Russell Munson
143: National Air and Space Museum
144: © Devaney Stock Photos
145: National Air and Space Museum
146: © Herman J. Kokojan/Black Star
146–147: © Robert DeGroat
148–149: © George Hall/Woodfin Camp & Assoc.
150 (both), 150–151: National Air and Space Museum
152–153: courtesy British Aerospace, Inc.
155: courtesy The Boeing Company
156: © Aram Gesar/Gesar Inc.
158: © 1981 James Sugar/Black Star
159: © Aram Gesar/Gesar Inc.
160, 161: © George Hall/Woodfin Camp & Assoc.
162–163, 164–165: © Herman J. Kokojan/Black Star
167: © Frederic Lewis
168–169: © James Sugar/Black Star
170–171: © Aram Gesar/Gesar Inc.
173: © 1981 James Sugar/Black Star
174 (top): courtesy The Boeing Company
174 (center): courtesy Musée de l'Air et de l'Espace

174 (bottom): 175 (all): NASA
176–177: © George Hall/Woodfin Camp & Assoc.
179: courtesy Musée de l'Air et de l'Espace
180: © Erik Simonsen
181: © Dan Moore
182–183: © Erik Simonsen
184: courtesy Dornier GmbH
185: © D. William Hamilton/The Image Bank
186: courtesy EAA Aviation Center
187: © 1988 George Hall/Woodfin Camp & Assoc.
189 (top left, top right): Howard Levy Collection
189 (bottom right): National Air and Space Museum
190: The Bettmann Archive
191: National Air and Space Museum
192: © Herman J. Kokojan/Black Star
193: © A. Carp/The Image Bank
194 (top, center): National Air and Space Museum
194 (bottom): R. Arnold Collection/National Air and Space Museum
195 (top): © Howard Levy
195 (bottom): National Air and Space Museum
196–197: © George Hall/Woodfin Camp & Assoc.
198: courtesy Northrop Corporation
199: courtesy Art Davis
200–201: National Air and Space Museum
202: Michael E. Long
203: courtesy Art Davis
204 (both): National Air and Space Museum
205: © James Sugar/Black Star
207: © Francis Thompson, Inc.
208: courtesy Art Davis
209, 210–211: courtesy EAA Aviation Center
212: © 1981 James Sugar/Black Star
214, 215: © 1986 Mike Fizer
216: © Erik Simonsen
217: Michael E. Long
218–219: © Herman J. Kokojan/Black Star
220 (left): © Erik Simonsen
220 (right): courtesy Grumman Corporation
221 (top): courtesy Messerschmitt-Bölkow-Blohm GmbH
221 (bottom): courtesy McDonnell Douglas Corporation
222: courtesy Northrop Corporation
223: courtesy Lockheed Corporation
224 (top): YF22 Team photograph, courtesy Lockheed Corporation
224 (bottom): courtesy Northrop Corporation
225 (top): Chuck Lanczkowski, McDonnell Aircraft Company, McDonnell Douglas Corporation
225 (bottom): courtesy Boeing Helicopter Company
226 (both): Komsolskaya Pravda

We have endeavored to obtain the necessary permission to reprint the photographs in this volume and to provide proper copyright acknowledgment. We welcome information on any oversight, which we will correct in subsequent printings.